Synthesizing Hope

Synthesizing Hope

Matter, Knowledge, and Place in South African Drug Discovery

ANNE POLLOCK

The University of Chicago Press
Chicago and London

The University of Chicago Press, Chicago 60637
The University of Chicago Press, Ltd., London

Published 2019
Printed in the United States of America

28 27 26 25 24 23 22 21 20 19   1 2 3 4 5

ISBN-13: 978-0-226-62904-9 (cloth)
ISBN-13: 978-0-226-62918-6 (paper)
ISBN-13: 978-0-226-62921-6 (e-book)
DOI: https://doi.org/10.7208/chicago/9780226629216.001.0001

Library of Congress Cataloging-in-Publication Data

Names: Pollock, Anne, 1975– author.
Title: Synthesizing hope : matter, knowledge, and place in South African drug discovery / Anne Pollock.
Description: Chicago ; London : The University of Chicago Press, 2019. | Includes bibliographical references and index.
Identifiers: LCCN 2018051002 | ISBN 9780226629049 (cloth : alk. paper) | ISBN 9780226629186 (pbk. : alk. paper) | ISBN 9780226629216 (e-book)
Subjects: LCSH: iThemba Pharmaceuticals. | Pharmaceutical industry—South Africa. | Drugs—Research—South Africa.
Classification: LCC HD9673.S64 I74 2019 | DDC 338.7/6161510968—dc23
LC record available at https://lccn.loc.gov/2018051002

♾ This paper meets the requirements of ANSI/NISO Z39.48–1992 (Permanence of Paper).

CONTENTS

INTRODUCTION

Hope in South African Drug Discovery

Pharmaceuticals are pleasingly tangible, tiny technoscientific objects that can be held in the hand, injected, or swallowed. At the same time, they are opaque representatives of a global pharmaceutical industry that is dauntingly complex.

Synthesizing Hope opens up the materials and processes of pharmaceuticals by starting in a distinctive place: iThemba Pharmaceuticals, a small South African start-up pharmaceutical company with an elite international scientific board, which was founded in 2009 with the mission of drug discovery for the treatment of tuberculosis (TB), human-immunodeficiency virus (HIV), and malaria. "iThemba" means "hope" in Zulu. iThemba was ultimately unsuccessful in finding new drugs, and the company closed its doors in 2015. Yet this particular place provides an entry point for exploring how the location of the *scientific knowledge* component of pharmaceuticals—in addition to their raw materials, production, licensing, and distribution—matters. I will explore why it matters for the scientists themselves and why it matters for those interested in global health and postcolonial science.

What if South Africa were to become a prominent place not just of raw materials, test subjects, and end users but of the basic science of pharmaceutical knowledge making? *Synthesizing Hope* is unusual in combining attention to global health and attention to postcolonial science, two spheres that are not often thought about together. In global health literature, scientists working in postcolonial contexts like Africa receive scant attention. Most global health research assumes that rich countries are the main, if not unique, source of knowledge making and that this knowledge flows "south." This exemplifies what anthropologists Jean Comaroff and John Comaroff have called the "epistemic scaffolding" of "Euromodernity," in which the West is the locus of "refined knowledge" and the rest of the world merely provides "reservoirs of raw fact."[1] *Synthesizing Hope* tracks a group of ambitious scientists' efforts to subvert that scaffolding.

In comparison with vast literatures on perspectives of patients and physicians, perspectives of pharmaceutical makers generally have been given insufficient attention.[2] There is a rich ethnographically engaged literature of global health research in Africa,[3] but not of pharmaceutical scientists there. This has some empirical justification: as anthropologist Kristin Peterson points out, the continent's pharmaceutical capacity has been in important ways "emptied out" in the processes of dispossession of biocapital.[4] More broadly, it is generally much more difficult for anthropologists and other social researchers to get access to pharmaceutical industry informants than to those positioned later in the pharmaceutical life cycle, such as prescribers and consumers.[5] The global pharmaceutical companies that make up "Big Pharma" are notoriously focused on controlling information,[6] and most start-up pharmaceutical companies follow suit. As a small company at the periphery of the global pharmaceutical industry, iThemba provided an intimate yet far-reaching perspective on preclinical pharmaceutical research, an understudied phase of pharmaceutical development.

Most scholarship of science in Africa has been of field sciences, agricultural sciences, and of labs that seek to isolate and verify field findings. In contrast, iThemba was a synthetic-chemistry-based company. The scientists there wanted to participate in global science as peers of elite scientists elsewhere. That is, they did not want to be limited to providing the Global North with problems to be solved, raw materials, or clinical trial subjects: rather, they wanted to participate in global knowledge creation. At the same time, the problems and possibilities they faced were rooted in their local context. Analysis of their project thus provides an opportunity for distinctive engagement with place.

South Africa is prominent in discussions of global health, but the country plays a peripheral role in conversations about global pharmaceutical science. Aspirations for scientific knowledge making are inextricably intertwined with infrastructures. The complex infrastructures used in most preclinical pharmaceutical research and development (R&D) today are highly geographically concentrated in what is variously called the "West" or the "Global North."[7] However, South Africa's specific history has left it with a scientific infrastructure that is much more developed than that in other parts of Africa, with well-regarded research universities, a robust electrical grid, and good transportation within the country and abroad. Just as Europe itself has always had "major centers, minor centers, and peripheries" of science,[8] Africa is not merely undifferentiated periphery: South Africa (Johannesburg in particular) is a node in networks of global science. iThemba itself was situated on the grounds of a historic dynamite factory—the largest in the world in its early twentieth-century heyday—and thus enjoyed robust and reliable access to electricity. This scientific and industrial infrastructure is the legacy of an oppressive history—colonialism and a system of legally enforced racial segregation and discrimination known as apartheid, which deprived the black-majority population of full citizenship. Now that South Africa is a couple of decades into an imperfect democracy, iThemba provides an opportunity to ask: could that infrastructure be turned to the service of the people?

iThemba scientists were inspired by the mission of creating new pharmaceutical treatments *by* and *for* South Africa and its region. They believed that it was important that this research be done in South Africa, and one common reason that they gave was that South Africa needed to recognize HIV (and TB) as its own problem and to take care of its problems itself. Self-reliance is in some senses an illusion: the small Johannesburg biotech company had an international board, as well as license agreements with international companies and universities, and anything developed there would also be part of global flows of knowledge production. But self-reliance was also an important component of how these chemists in South Africa understood the terrain of raising the profile of their scientific community and their country.

Post-apartheid South Africa is an illuminating site for analysis of the allure and the difficulty of the creation of *science* (universally applicable knowledge) that is *democratic* (accountable to particular publics). If, as historians and science and technology studies (STS) scholars Paul Edwards and Gabrielle Hecht argue, nuclear and computer systems were key in the technopolitics of apartheid, the iThemba project might be conceptualized as a gesture toward a potential technopolitics of post-apartheid.[9] For Edwards and Hecht, computer systems functioned as a tool and as a symbol of apartheid South Africa, both within the country and as a focal point for outsiders. iThemba's project of drug discovery for TB, HIV, and malaria can be understood to play an analogous role for a democratic South Africa, albeit in a limited way in light of that effort's small scale. For both South African political leadership and the range of national and international actors who came together to found iThemba, science and technology generally, and drug discovery science in particular, became sites for nation building.

Since iThemba means "hope," how to account for its eventual failure? The small company might be understood to play the role of the hero of a tragic narrative, perhaps destined from the beginning to experience downfall in the face of the overpowering forces of South African and global social, political, and economic orders. iThemba's ultimate failure to find new drugs or even a sustainable business model points toward the difficulty of sustaining both a small drug company and large nationalist aspirations, even as the hopes that iThemba brought together live on in other ways. My account seeks to illuminate both iThemba's ambitions and the intransigence of social, political, and economic orders. Even after its failure, however, iThemba continues to offer a place of hope.

Introducing iThemba

Since the name "iThemba" means "hope" in Zulu, it is perhaps fitting that most of what I will describe in this book are aspirations rather than actualities. iThemba was initially founded around 2000, quickly floundered, relaunched in 2009, and closed its doors in 2015 (fig. 1). I did my research over the period of 2010–15. Reference to "hope" is consistent with R&D generally—a colleague in accounting at my university in the United States likes to say that "R&D stands for hope," which is to say that when com-

1. iThemba Pharmaceuticals. Photograph by Katherine Behar.

panies assign a valuation to R&D, they are assigning a value to hope. Although the company did make incrementally important contributions to knowledge that live on in several patents and publications, it ultimately did not make any drugs. Indeed, it did not make much of its own: a large portion of the scientists' time was spent generating revenue by synthesizing molecules on contract for pharmaceutical companies elsewhere. iThemba's drug discovery goals were ultimately unrealized. iThemba's project was aspirational more than actual, yet this book shows that the aspirations themselves are worthy of consideration.

Over the years, the name of the company became a bit confusing for many international audiences. iThemba does not have the same etymology as iPads and the like but, rather, a Zulu etymology. Zulu is the most widely spoken language in South Africa, with about a quarter of South Africans speaking it at home and more than half able to understand it. In Zulu (or isiZulu, as it's called in Zulu), nouns consist of a prefix and a noun stem: in this case, iThemba. The *th* is not the sound in "think" or in "this" but instead an aspirated *t* like the one in "tin" (in the breathy way that most English dialects, including American English and British English, pronounce an initial *t* followed by a vowel). The word "iThemba" is a very common one in South Africa that is widely known to people there no matter what home language they speak and whether or not they speak Zulu. Like Spanish words such as *por favor* and *gracias* in the United States, someone in South Africa does not need to know Zulu to know the meaning of "iThemba." That said, the name iThemba Pharmaceuticals was also confusing for South African audiences, because iThemba is a common name for nonprofits and other initiatives, and there is even a company in Johannesburg called iThemba Labs. Yet despite the frustrations that came with the ambiguity and confusion of the name, most of the members of the diverse team of scientists involved thought that "Zulu for hope" was a good way to describe their endeavor.

Synthesizing Hope draws extensively on the perspectives of iThemba's drug discovery scientists, who were composed of two groups: members of the company's management and Scientific Advisory Board, who were internationally trained and based in the United States, United Kingdom, Switzerland, and South Africa; and bench scientists, who were trained in South Africa and worked in Johannesburg. Since beginning my research on this project in 2010, I have interviewed several members of iThemba's management and Scientific Advisory Board—some in the United States and the United Kingdom, some in South Africa—and I have taken several research trips to iThemba's lab. All the scientists at iThemba agreed to participate in multiple open-ended interviews, and I also attended their lab meetings and did participant observation on-site. My participation was more circumscribed than is canonical in anthropological research, since I participated only in lab conversations of various kinds and not, for example, in hands-on lab work side by side with the bench scientists, but the intimate group of chemists did warmly incorporate me into their lab's social world.[10]

A small corporate scientific setting poses challenges and opportunities for ethnographic methods. I originally gained access to iThemba as a field site through its Emory University–based cofounder and Scientific Advisory Board member, Dennis Liotta—an extraordinarily successful drug discovery scientist who had discovered key antiretroviral drugs that transformed HIV/AIDS from an inevitably deadly disease to a potentially manageable chronic one.[11] I first heard Liotta speak about iThemba in 2008 and first met him in 2010. I had e-mailed Liotta cold with a request to meet to find out more about iThemba and was pleasantly surprised by his openness. As a junior faculty member at a respected university, I found my positionality to be somewhere between that of a colleague and a mentee. As someone much younger and less established in my career than Liotta, in a field with considerably less prestige, and as a woman seeking to study a male-dominated scientific field, I was highly aware of the fact that in reaching out to him I was "studying up."[12] However, my PhD from MIT and faculty status at Georgia Tech gave me credibility as a domain expert in anthropology of science. Liotta seemed genuinely happy to bring me into the fold of his drug discovery circles. And as he introduced me to his scientific colleagues from South Africa, I was also conscious that I shared with Liotta privileged status as a white American. In my engagement with drug discovery scientists based in South Africa, power inequalities between natural sciences and social sciences were in tension with those that ran in the opposite direction, along colonial lines.[13]

Liotta introduced me to several members of the company's executive leadership and advisory board at a dinner in a private dining room of the W Hotel Atlanta Downtown in early May 2010. These dinner companions were all in Atlanta for a short visit, to attend a meeting. When I went to Johannesburg for the first time just a few weeks later, those executive leaders were generous hosts, allowing me extraordinary access to the company. For example, they allowed me to sit in on confidential meetings and to interview scientists on-site and on company time. However, in light of my close connections with their bosses and the small size of the company—at that point, about a dozen bench scientists—the bench scientists were taking some risks by being candid in their conversations with me. I assured them that I would protect their privacy by not making them individually identifiable in the text that I produced. This decision, in turn, ruled out the writing of a particular kind of intimate ethnography and created the conditions for a somewhat more distant account.

Synthesizing Hope puts these interviews and observations into the context of a much larger history: of colonial models in which Africa's role in the global economy was predominantly as a site of extraction of raw materials; of apartheid and struggles against it; of global health advocacy since the emergence of HIV/AIDS; of a generation of scientists who have come of age in an era defined by all these intertwined histories; and of a historical moment of change in synthetic chemistry and the global pharmaceutical industry. The book also puts the company's project into the context of the contemporary events happening over the course of my research: my first visit was during the lead-up to the men's 2010 FIFA World Cup (see fig. 2), a time of tremendous optimism, but the period of my research was also marked by political frustrations and by the death of Nelson Mandela.

The founding of iThemba was a collaboration between the international board of cofounders and South African government funders. The international scientists contributed intellectual property and other intangible resources. The South African government provided start-up funds through its Technology Innovation Agency, part of the Department of Science and Technology, in exchange for a 50.1 percent stake in the company. That the company received those start-up funds from a science and technology

2. The reception desk at iThemba: staff clothing and figurine in honor of the 2010 World Cup. Photograph by the author.

department rather than from a health department would contribute to particular challenges. For example, it put the company in the position of competing for funding with prestige projects in physics and with information technology ventures that have much shorter timelines for return on investment than do those in biotechnology.

iThemba selected molecules for drug discovery research in a variety of ways, sometimes in collaboration with universities doing high-throughput screening of large numbers of potential compounds[14] and sometimes in a more focused way, proceeding on the basis of knowledge of the target proteins and literature about classes of chemicals that are likely to have activity at that target.[15] Some of this research drew on "patent pools" from universities and Big Pharma companies, which were designed to share information among those pursuing novel therapies for neglected diseases.[16] These were all sources of what are referred to as "hits": small molecules that seem to be active against the target (in iThemba's case, TB, HIV, or malaria).

For anything chosen for further development, the next step was to validate the hits—confirm the activity, specificity, and selectivity of the molecules. For example, the molecule should be toxic to the target pathogen but not to relevant tissues grown in culture. In addition to generating experimental data, the scientists also searched specialized databases to make sure that there was freedom to operate in terms of intellectual property; that is, a molecule developed from the hit should be potentially patentable.

Then, the scientists sought to turn the "hits" into "leads" by improving the potency, selectivity, and physiochemical properties of the molecules of interest. Many analogues would be synthesized and assessed. The specifics of iThemba's lead molecules and approach to optimizing them were secret while under way, though some of the exploration of potential anti-TB compounds in their library has since been published.[17] iThemba's Scientific Advisory Board members have discovered successful drugs through these types of research programs at their universities and at other companies they have been involved with, and it could not have been known in advance that iThemba would not have that same success.

The business model of iThemba Pharmaceuticals changed over time. The need to generate revenue while doing drug discovery work was always a challenge for the small company. To generate cash flow and build scientific skills while working on long-term drug discovery projects, iThemba took on contract chemistry projects, synthesizing molecules for other pharmaceutical companies in Europe and elsewhere. It also did contract work for major international initiatives, including the Drugs for Neglected Diseases Initiative and the Medicines for Malaria Venture. Under pressure from gov-

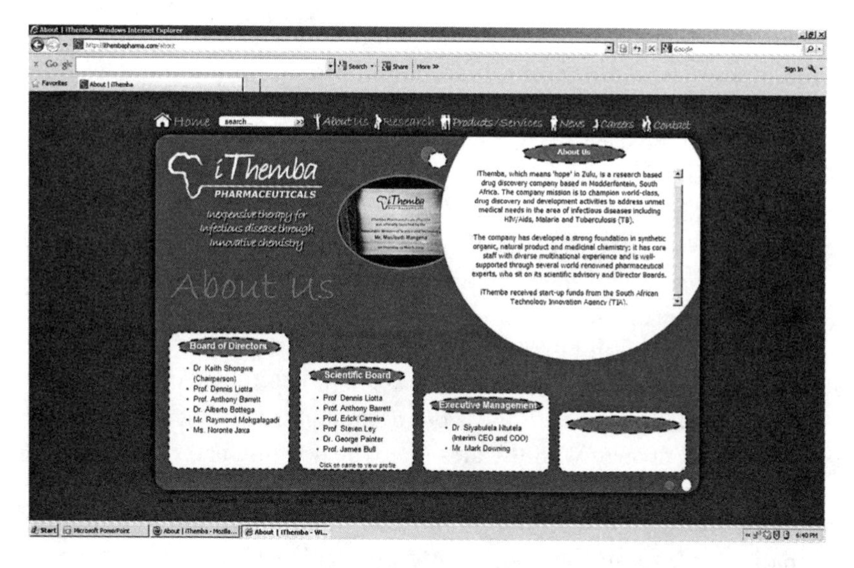

3. The website of iThemba Pharmaceuticals at the company's height: ithembapharma.com, accessed August 8, 2013.

ernment funders to make something useful in the shorter term, the company later sought to build a "green" pharmaceutical manufacturing capacity that would both manufacture antiretrovirals for the local market and be well positioned globally for future demand for greener manufacturing processes. Ultimately, the operating cash flow from contract chemistry was not enough of a supplement to the South African government's support to make the enterprise sustainable, and the capital investment required for green manufacturing was not forthcoming either.

The main web page for iThemba during its heyday (fig. 3) can provide one route into analyzing the company's project. The website was addressed to multiple audiences: South African bureaucrats and scientists and plural international publics. The web page invoked its "world renowned" advisers, and for an audience in-the-know, the listed names highlighted the global prestige of the company: the North American– and European-based scientists on its boards included some of the most elite drug discovery scientists in the world. But the page also highlighted the South Africanness of the project: the Zulu name of the company, its focus on diseases for which treatment is urgent in Africa, and its support from the South African government through the provision of start-up funds. The logo and the graphic design of the site as a whole were developed by a designer in the United Kingdom, but there is an informality to the page, with prominent use of italics that might evoke handwriting, and dotted lines used as borders for text boxes that might evoke handicraft.[18] The design as a whole projects a rather modest enterprise, even as the text and the names of the board of directors and the scientific board listed project power.

The most interesting aspect of the web page is the logo. The logo on the website appears to be a copy of the more professionally designed logo on-site, which appears in a small photograph of a sign in the center of the page. The logo of the company incorporates place, featuring a simplified, stylized outline of a map of Africa in which the *i* of "iThemba" replaces the right-hand border, and the rest of the word extends beyond where Africa's geographic border would be. The removed right-hand border of the map of Africa becomes a door that opens out to the right. The visual language of the logo is consistent with the idea that iThemba originates in Africa and extends out into the world.

A central hope that motivated those involved was the idea of creating capacity for "African solutions for African problems," by doing synthetic-chemistry research in South Africa to find new treatments for diseases prevalent in Africa. "Africa" is a fraught referent, rooted in colonial ideas and ongoing epistemological and material inequalities,[19] and South Africa is a particularly problematic stand-in for the continent as a whole. Yet the term's capaciousness does some important work. "Africa" is strikingly flexible, able to incorporate and inspire South Africans of diverse ethnicities, as well as (black) Africans from other parts of the continent—a rhetorical move that has a long history in antiracist politics in South Africa.[20] Moreover, a wide range of social actors in South Africa and elsewhere on the continent draw on the idea of Africa to make sense of their situations and construct strategies for improving them.[21]

Matter, Knowledge, and Place

iThemba provides a useful starting point for distinctive engagement with the materiality of pharmaceuticals and pharmaceutical infrastructure, from the perspective of their (would-be) makers. iThemba was first and foremost a drug discovery company, which means that its priority was making intellectual property—patentable molecules—rather than making actual pharmaceuticals for distribution to consumers. And yet, scientists there frequently highlighted connections between pharmaceutical knowledge and pharmaceutical stuff.

Pharmaceutical knowledge can seem placeless when it is abstracted into a patent, but it is necessarily created in particular times and places.

iThemba's work to create globally respected pharmaceutical knowledge was not unconnected to its local context. A few ultimately unrealized endeavors can help us to see how.

For example, the form of a drug as a pill or liquid or gel may not seem all that important, but it can be—not only for consumers but also for producers. The period in which iThemba was operating was one in which there was a lot of excitement about the potential for antiviral gels that women could use vaginally to prevent HIV transmission—hopes that have since been largely dashed by failed clinical trials. For health practitioners and activists, these gels were held up as emblems of women's empowerment in the fight against HIV.[22] For pharmaceutical makers, there was another appealing element to the gels, besides just a new product to market: gels contain a lower percentage of active pharmaceutical ingredients than pills do, and so local production of gels seemed more feasible than local production of pills. Active pharmaceutical ingredients—APIs—are a key part of pharmaceuticals, not only in terms of efficacy but also in terms of the simple ingredient composition. South Africa lacks capacity to make APIs, which means that companies there must import this key component in order to formulate drugs. Decreasing the amount of an API required could make local manufacture of the API more realistic or, at least, decrease the burdens and risks of relying on imports. The material composition of drugs matters.

Frequency of dosing is another example of a material aspect of drugs that matters. In the HIV/AIDS treatment space, this is most often understood in terms of "pill burden": patients have an easier time complying with once-a-day, single-pill regimens than with complex, multi-dose, multi-pill regimens, which leads to increased adherence to treatment and decreased viral load.[23] But simplicity from the patient's perspective is just one element of how dosing matters. Consider the common instruction on pharmaceutical packets to "take *x* times daily with food." The scientists at iThemba helped me to see how this imperative could be insensitive and even cruel. Hunger is a condition that is rarely discussed in the literature of clinical trials, despite its pervasiveness and its effect on treatment efficacy,[24] but iThemba scientists were highly aware of its role in the lives of those for whom they hoped to develop treatments.

This awareness is one reason that the scientists were excited about an idea that was not ultimately pursued but was plausible given the technologies that the company had licensed: the prospect of once-a-week treatment for HIV. The South African government procures most of the antiretrovirals for the population, and once-a-week treatment would lower its cost of pills:

packaging costs would be lower, pill-production costs would be lower, distribution costs would be lower. But a once-a-week pill would also help to lower the costs borne directly by the very poor people in need of antiretrovirals: they would need transport to the clinic less frequently. One scientist went on at length about the importance of increasing employment levels for effectively expanding HIV treatment, because patients need access to income. At the same time, the scientist suggested that a once-a-week pill also offered potential amelioration in the meantime: patients would be able to take these drugs even if they had access to only one proper meal per week.

A third example of how pharmaceuticals' materiality matters is an endeavor that made it further along in development, such that I devote a whole chapter to it: "Hope in Flow" (chapter 6). Pharmaceutical consumers generally don't think much about the material processes that preceded the products' packaging, but we should. One reason is that pharmaceuticals have a disproportionately high environmental impact relative to other industrially manufactured products, and most of that impact happens at the production stage. Although, for example, fossil fuel industries produce much more total waste, the standard manufacturing processes for pharmaceuticals produce a much higher proportion of waste per ton of product. iThemba was operating during a period in which there was interest, including among those setting the policy agenda in the European Union, in taking the environmental impact of products into account no matter where that impact was borne. Many pharmaceuticals generally and antiretrovirals in particular are purchased by a relatively small group of payers—governments and nongovernmental bodies of various kinds—and if those payers were to become willing to pay a premium for pharmaceuticals produced in a "greener" way, that would create a potential market niche for those able to provide greener processing. In its continuous-flow chemistry project, iThemba hoped to leapfrog standard toxic manufacturing processes and build green manufacturing capacity.

All three of these unrealized innovations—antiviral gels, once-a-week antiretrovirals, and green pharmaceutical manufacture—are recompositions of matter that also sought to be interventions into sociopolitical contexts. They would have negotiated intellectual property in different ways—reformulating off-patent molecules, developing patentable molecules, and developing trade-secret manufacturing processes—as they aimed to intervene in distinctive elements of context. But both matter and sociopolitical contexts are hard to change, and great ideas like these turn out to be extraordinarily difficult to realize.

All these elements of the materiality of pharmaceuticals point to the important role of infrastructure. Pharmaceuticals have long traveled the globe, but pharmaceutical knowledge making has become concentrated in just a few geographic places. Pills themselves are highly mobile, and the need to increase the reach of essential medicines has become a priority for improving global health. But the science of pharmaceutical knowledge making, especially at its most fundamental stages, is not so mobile or dispersed. Even pharmaceutical manufacture is notably concentrated, as "the geographies of pharmaceutical infrastructure" become located in the specific sites in which "good manufacturing practices (GMP) can be achieved."[25] In accounts of clinical practice in Africa, scarcity of laboratory capacity is often a prominent feature of Africa's incomplete incorporation into biomedicine as practiced elsewhere.[26] Yet relative to other parts of the continent, South Africa has a great deal of capacity. These ties between infrastructure and pharmaceutical knowledge making are a central theme of this book.

A Local and Global Story

iThemba provides an opportunity to tell a simultaneously local and global story or, perhaps, a local story of aspirations to the global. As anthropologist Anna Tsing has argued, "Capitalism, science, and politics all depend on global connections. Each spreads through aspirations to fulfill *universal* dreams and schemes. Yet this is a particular kind of universality: It can only be charged and enacted in the sticky materiality of practical encounters."[27] The scientists at iThemba were participating in a large global project of drug discovery, but doing so from a particular, small South African lab. Historians of colonial science David Wade Chambers and Richard Gillespie point out, "The local and the global are a dialectical pair and must remain so in our histories."[28] To understand center/periphery relationships in global research collaboration today, dichotomous understandings continuous with colonialist discourse are necessary but not sufficient. At iThemba, the global legitimized the local, and vice versa, such that the global and local grew at the same time.

There are many ways to characterize the most fundamental divide across which scientists at iThemba were working—between rich countries and poor ones, between the first world and the third world, between the metropole and the periphery, between the Global North and the Global South. The terms Global North and Global South are a bit confusing as applied to South Africa, because although South Africa is at the southernmost tip of Africa geographically, it is the African country closest to the Global North in terms of economic development. Yet Global North and Global South are the terms that I tend to use when talking about global health, because they are more current in that discourse than alternatives such as the hierarchical first world / third world or the teleological developed/developing. Center/periphery also constitute a prominent pair of poles in this book, especially when talking about the social studies of science, because those terms are particularly useful for thinking about networks. Suffice it to say, all these binary mappings are problematic and incomplete—yet they are relevant nonetheless.

Analysis of iThemba problematizes Global North/South divides, illustrating ways in which the worlds of scientists are not so neatly bifurcated, at the same time that it highlights how much is at stake in who makes knowledge and where. Global North/South bifurcation is necessarily slippery terminology, because although it usefully captures the role of geography and colonial history in structuring development, it can obscure the mobility of scientists across the divide and the heterogeneity within the South—since countries like South Africa are "home to both world-class research institutions and understaffed ill-equipped" ones.[29] Amelioration of disparities between South Africa's own historically white universities and historically black universities brings broader, global health capacity-building initiatives to a within-country scale.[30]

South Africa is an interesting intermediary in Global North/South bifurcation because its particular postcolonial context is shaped by its legacy of settler colonialism, apartheid, and the struggle against apartheid. South Africa is a relatively young democracy, resource poor by global standards, with a large population of impoverished black Africans with urgent unmet health needs, and with robust activist communities that have mobilized around demands for access to medicine. Thus, it is an obvious locale for the interventions that tend to travel under the rubric of "global health." At the same time, South Africa is highly developed by African standards, with a rich history of innovation in biomedicine—such as the world's first human heart transplant, in 1967—and a now-multiracial middle class with the education, resources, and global connections to plausibly participate in global science. In contemporary aspirations for South African science, a potential exists for those two South Africas to come together through science in the service of the people.

Even as science and technology have been part of articulating South Africanness since the colonial period, there are important ruptures. In the early twentieth century, there were aspirations that science could unite white South Africans across ethnic divides as they contributed to the global "commonwealth of knowledge."[31] During the apartheid era in the mid-twentieth century, there was significant innovative science in South Africa, but it was more globally isolated and for the benefit of the country's white minority. By the early twenty-first century, South African science had become at once more inclusive nationally and regionally but more marginal globally. The transition from the colonial and apartheid models to one in which science would serve (the majority of) the people has been a rocky one—and is incomplete.

South African Drug Discovery as a Sociotechnical Imaginary

It is not clear whether the hope for science in the service of the people will be realized in some element of the afterlives of iThemba Pharmaceuticals. Science and technology in South Africa remain dominated by extraction industries, and fledgling efforts to foster drug discovery there are highly tenuous. Nevertheless, tracking these efforts provides insights for STS and beyond. As Sheila Jasanoff and Sang-Hyun Kim have argued, imagined possibilities matter; what they describe as "sociotechnical imaginaries" are vital sites for thinking through not just what technology should be but also what the nation should be: "technoscientific imaginaries are simultaneously also 'social imaginaries,' encoding collective visions of the good society."[32] Although Jasanoff and Kim had originally articulated the concept of sociotechnical imaginaries at the level of the nation-state, Jasanoff later extended it: "Sociotechnical imaginaries . . . can be articulated and propagated by other organized groups, such as corporations, social movements, and professional societies. Though collectively held, sociotechnical imaginaries can originate in the visions of individuals or small collectives, gaining traction through blatant exercises of power or sustained acts of coalition building."[33] iThemba might be a questionable case for the concept from Jasanoff's perspective, since she argues that "only when the originator's 'vanguard vision' . . . comes to be communally adopted does it rise to the status of an imaginary."[34] iThemba successfully materialized to a nontrivial degree, becoming the site of commitment and energy from a network of international scientists, successfully gaining funding from the South African government, and existing as a space of bench scientific work over the course of several years. To at least some degree, in Jasanoff's terms, "key actors" were able to "mobilize the resources to make their visions durable."[35] That said, the iThemba project was certainly not widely known about in South Africa or anywhere else, and it was not the site of social movements or other broader mobilization.

Nevertheless, I find the concept of sociotechnical imaginaries to be useful. As in Jasanoff's cases, analysis of iThemba in these terms allows exploration of "the normativity of the imagination and the materiality of networks."[36] "Normativity" here denotes claims about what should be done, and "materiality" points to efforts to realize those aspirations in space and time. As I will explore throughout this book, a sociotechnical imaginary organized around locally discovered, innovative drugs for infectious diseases articulates South African scientific research and South African society in ways distinct from a range of alternative available imaginaries—of aid complexes or plant-based therapies; of extraction industries; or of colonial- or apartheid-era science in which African populations were not seen as the principal beneficiaries of biomedical innovation. Other scholars have articulated South African histories and presents through other particular state-supported high-tech projects, including computing and nuclear weapons,[37] liquid fuel made from coal,[38] and biometrics.[39] What might the particular pharmaceutical endeavor of iThemba offer?

Drug discovery is a particularly interesting site for analysis of these sociotechnical imaginaries because pharmaceuticals are so readily recognized as the material-semiotic, in which it is possible for objects to simultaneously carry matter and meaning.[40] Pharmaceutical science can be a site of reconfiguring facts and social orders.[41] For consumers, all drugs are "recreational," since they are means of re-creating ourselves and our worlds,[42] and for scientists they can be too. Anthropologists of pharmaceuticals have explored how pharmaceuticals can help to "make dis-ease concrete" for both patients and physicians,[43] and here we can see how they can help to make hope concrete for synthetic chemists.

In many conversations that I had during my time with South African drug discovery scientists, I was often struck by the materiality of pharmaceuticals: not simply thinginess as a concept[44] but how the stuff of pharmaceuticals matters. As much as anthropologists and STS scholars have theorized the ways that bodies matter both in terms of meaning and in terms of materiality (in Judith Butler's classic formulation),[45] few have paid much attention to drugs' material consequences for bodies,[46] much less drugs' own materiality. But for their developers, the materiality of drugs is inescapable. And South African would-be developers face particular material constraints. Costs of packaging and moving around these tiny objects are nontrivial in this context. And the layers of production have costs too.

The costs of production and distribution highlight the complexity of the terrain of neglected diseases. Two of the three diseases that iThemba researched—malaria and TB—are at least plausible cases for the canonical definition of neglected disease (albeit less so than diseases that are more geographically limited). But context-relevant HIV treatment—that is, new-generation HIV treatment that is easier to distribute and lower cost—can be understood as neglected too: it is also something not profitable for Big Pharma but for which there is a tremendous need. This context is part of the situatedness of this South African pharmaceutical company.

Of course, there are other reasons that it matters that pharmaceuticals should be discovered and made in South Africa, beyond materiality. A lot of meaning is wrapped up in that situatedness as well. Anthropologist Susan Whyte and colleagues were talking about patients and clinicians when they observed that "pharmaceuticals do hermeneutic (as well as pharmaceutical) work,"[47] and this applies to drug discovery scientists as well. Pharmaceuticals have the power to re/define illness and treatment,[48] and they do hermeneutic work for their discoverers and developers as well as for their patients.

South African drug discovery becomes another layer of what anthropologist João Biehl has termed pharmaceuticalization.[49] In Biehl's attention to the pharmaceuticalization of the Brazilian response to HIV, he is particularly attentive to the way that public health has been pharmaceuticalized. As anthropologists elsewhere have pointed out, too,[50] promoting affordable pharmaceutical access as the solution to dire health problems renders a very narrow concept of the scope of public health. When access to pills stands in for access to health, it continues systematic exclusions. Pharmaceuticalization critiques highlight that pharmaceuticals are not sufficient solutions to problems of ill health, much less global inequality, but that does not necessarily mean that pharmaceuticals cannot be part of those solutions. In efforts to create drug discovery capacity in the Global South, there is a tension between aspirations of repurposing tools developed by and for global capitalism for the benefit of the marginalized people of the world and a sense that "the master's tools will never dismantle the master's house."[51]

An African pharmaceutical response is particularly charged in post-apartheid South Africa, a context in which ideologies of neoliberalism, democracy, HIV denialism, and pharmaceutical-based activism all intertwine and contest each other.[52] The pharmaceuticalization of public health is part of what is going on in this South African case, but there are other things being pharmaceuticalized as well. By looking not only at the distribution of pills but also at their design, we can see other aspects of pharmaceuticalization. For the iThemba endeavor, locally discovered, affordable pharmaceuticals were the answer to the problem of public health in South Africa—or at least part of the answer. But pharmaceuticals also become the solutions to many more problems: that of transforming an extraction economy into a knowledge economy, stemming brain drain, and raising the profile of the country. In this start-up, South African science was itself being pharmaceuticalized. When not just objects of pharmaceutical knowledge but also pharmaceutical knowledge production itself moves to the Global South, we have an opportunity to explore implications for STS and beyond.

iThemba offers an interesting opportunity to take on Bruno Latour's classic STS challenge, expressed in his essay "Give Me a Laboratory and I Will Raise the World," to eschew both internalist and externalist accounts of science. Even at that early stage of his writing, Latour was drawing attention to the tenuousness of the capacity of laboratory work to gain the social support that it requires to happen in the first place and then to get its products out into the world: "People readily give their attention to someone who claims that he has the solution to their problems but are quick to take it back."[53] Drawing on Latour's terms, I might suggest that iThemba was able to construct a provisionally compelling translation of the problems of infectious diseases in (South) Africa, but that translation did not endure and was thus insufficiently realized.

My analysis of iThemba, an ultimately incompletely realized technological project, has perhaps deeper resonances with Latour's more imaginative text *Aramis; or, The Love of Technology*. For Latour, that never-built Paris public transportation system provides an opportunity to explore the contingent possibilities of a technological project. As Latour points out, "by definition, a technological project is a fiction, since at the outset it does not exist," and in order to be realized, the technological project must successfully bring on board an unruly collection of human and nonhuman actors that are always at risk of losing interest.[54] As with Aramis, no single person or thing is responsible for iThemba's ultimate failure—in the end, the network did not hold.

Synthesis and Hope

The title of this book, *Synthesizing Hope*, can be unpacked into intertwined themes of synthesis and hope. Consider the etymology of "synthesis": via Latin from the Greek *sunthesis*, from *suntithenai*, "to place together." Thus, synthesis provides an opportunity to foreground the copresence in space of the intertwined materialized histories of global health and global pharma, of colonization and mining, of apartheid and post-apartheid: place matters.

The histories that I have mentioned here are also sets of tensions, and so synthesis has other relevant connotations. The Hegelian sense of synthesis refers to the dialectical method that highlights the productivity of contradictions: proposition (thesis), opposition (antithesis), and new proposition (synthesis).[55] In Marxist dialectical thinking, Hegelian thought has offered ways to think about history and struggle.[56] When we apply it to the case at hand, we might understand the desire to make inexpensive innovative drugs for infectious diseases in South Africa as a synthesis of the contradiction between Big Pharma's intellectual property regimes and access-to-medicines campaigns. We might understand post-apartheid scientific capacity as a synthesis of apartheid and the struggle against it. And we might understand South African chemistry as a synthesis of an exploitative economy dominated by extraction and dreams of a new kind of knowledge economy. Considering the plural, intertwined histories in turn resonates with the Kantian sense of synthesis: combining and unifying isolated data into a comprehensible whole.

Far less erudite senses of "synthetic" are at play here as well. iThemba was a synthetic-chemistry-based company, and so synthesis was fundamental to their work in the sense that they were combining substances to produce a desired product. And the perhaps most prosaic sense of "synthetic" will become important as well: "synthetic" meaning artificially produced, nonnatural. This points to a contrast between iThemba's approach and other routes to pharmaceutical innovation more widely associated with Africa that rely on bioprospecting and traditional knowledges.

Turning now to hope: "hope" is a common word, and yet a complicated concept. As Cheryl Mattingly argues, hope can be usefully understood "as a *practice*, rather than simply an emotion or a cultural attitude."[57] The low chance of success of iThemba's drug discovery goals also resonates with her observation that "[p]aradoxically, hope is on intimate terms with despair. It asks for more than life promises."[58]

Hope is an engine on which biocapital operates,[59] but it is also something more than its entrepreneurial elements. It is an essential part of postcolonial social movements. As the educator and philosopher Paulo Freire argues:

> I am hopeful not out of mere stubbornness, but out of an existential, concrete imperative. I do not mean that, because I am hopeful, I attribute to this hope of mine the power to transform reality all by itself, so that I set out for the fray without taking account of concrete, material data, declaring "My

> hope is enough!" No, my hope is necessary, but it is not enough. Alone, it does not win. But without it, my struggle will be weak and wobbly. We need critical hope the way a fish needs unpolluted water.[60]

Many in STS and related fields have described the ways that pharmaceuticals become a site of hope, for advocates and consumers and correspondingly for marketers.[61] Anthropologist Joe Dumit has insightfully unpacked the work done when direct-to-consumer advertisements intone that "there is hope."[62] *Synthesizing Hope* tracks how pharmaceuticals become sites of hope for those who fabricate them as well. The iThemba scientists were not activists per se, but they were trying to change the world.

Hope had a great deal of international salience during the period of iThemba Pharmaceuticals' short life, because it was also the era of US president Barack Obama. In the 2004 speech that catapulted Obama to fame, he endorsed the candidates John Kerry and John Edwards by invoking hope in a way that was very resonant with that of Freire: "I'm not talking about blind optimism here. . . . No, I'm talking about something more substantial. It's the hope of slaves sitting around a fire singing freedom songs; the hope of immigrants setting out for distant shores; . . . the hope of a skinny kid with a funny name who believes that America has a place for him, too. The audacity of hope!"[63] As I write this after the failure of iThemba and the end of Barack Obama's presidency, hope seems more elusive, yet no less essential.

Chapter Outlines

Although the book as a whole has an argument and narrative arc that transcend chapters, individual chapters can also be read on their own. Readers with particular interests might select particular chapters—for example, chapters 1 and 4 for global health; chapters 2, 5, and 6 for chemistry; chapters 2, 3, and 4 for South Africa. Some of the chapters are more concrete and so more accessible to readers with less familiarity with the scholarly literatures engaged, especially chapters 2 and 4. Graduate students and others interested in theoretical concerns will find those to be most centrally engaged in chapter 5.

The first chapter, "Questioning the Bifurcations in Global Health Discourses," delineates the distinctiveness of iThemba's project of drug discovery for infectious diseases in South Africa by contrasting it with three more common sets of discourses about pharmaceuticals in South Africa: access-to-medicines campaigns; traditional knowledges and bioprospecting; and clinical trials. It argues that these three rely on a bifurcation between North and South, which analysis of iThemba helps to trouble.

The second chapter, "In the Shadows of the Dynamite Factory," introduces the physical site of iThemba Pharmaceuticals in a place called Modderfontein, on the edge of the grounds of a historic Nobel dynamite factory, in order to explore the material and social legacies of extraction industries for the possibilities of and constraints on drug discovery and manufacture in South Africa. It analyzes parallels between dynamite and pharmaceuticals and the hopes for building on a technical legacy while moving forward from a violent colonial and neocolonial past.

The third chapter, "Science for a Post-apartheid South Africa," focuses on a related legacy. Drawing on accounts of iThemba's cofounding scientists and executive leadership, it explores how iThemba became a site for both local and international investment in the hope for a democratic South Africa. Apartheid legacies create both challenges for realizing this goal and conditions of possibility. In the new South Africa, science should serve the people, but the neoliberal era presents obstacles for building science.

The fourth chapter, "African Solutions for African Problems," turns to the hopes that the bench scientists at iThemba had of participating in global science, not just as providers of raw material or recipients of end products but as participants in pharmaceutical research. Although the synthetic-chemistry work that they did could conceivably be done in any well-equipped lab in the world, they articulated drug discovery in South Africa as rooted in place in three ways: connected with personal experiences of disease; an expression of commitment to democratic citizenship; and providing the possibility of working "at home." Like the perspectives of scientists everywhere, theirs were rooted in their context. Unlike scientists in the Global North, however, they were highly aware of that situatedness.

The fifth chapter, "Im/materiality of Pharmaceutical Knowledge Making," explores the challenges of building a "knowledge economy" through the necessarily material route of pharmaceutical discovery and development. Intellectual property can exist in abstract forms, but in order to become pharmaceuticals, it must be materialized with products and processes that are unevenly distributed in space. In iThemba's work, the importance of "the map" is both palpable and reconfigured.

The sixth chapter, "Hope in Flow," explores a new direction that the company had hoped to follow in order to fund its discovery work: building "green" pharmaceutical manufacturing capacity in South Africa by implementing continuous-flow chemistry. This provides an opportunity to reflect on time and materiality, both on the microlevel of the chemical reactions that are carried out to synthesize molecules and on the macro-level of providing a route for South Africa to become a site of pharmaceutical innovation. In a way analogous to how cell phones have leapfrogged landlines in Africa, the lack of pharmaceutical manufacturing capacity becomes a condition of possibility for skipping to the next paradigm in drug manufacture.

The epilogue, "The Afterlives of Hope," describes plural legacies of iThemba's project. Even though iThemba was unsuccessful in discovering new drugs for TB, HIV, and malaria, it was generative. Its continuing generativity can be seen in the paths taken by the scientists and scientific materials that have moved on from iThemba. The epilogue also outlines some of the conceptual legacies of the book for STS, anthropology, and global health research. It reflects on iThemba's tragic narrative arc to consider how this small company's experience can help to illuminate larger social, political, and economic orders.

ONE

Questioning the Bifurcations in Global Health Discourses

In October 2008 I was in my first semester on the faculty at Georgia Tech and keen to meet other scholars and activists in Atlanta with whom I might share interests. One afternoon, I took the bus to nearby Emory University to attend the "International Access to Medicines" panel of a human rights conference held at the School of Law.[1] It was a decision that I made on a whim, but it would turn out to be a fateful one. That panel was where I first learned about the small South African pharmaceutical company that would become my field site: iThemba Pharmaceuticals.

The panel focused on access to medicines in poor countries. One of the speakers was an Emory chemistry professor, Dennis Liotta, a prominent drug discovery scientist. Liotta and his collaborators were the researchers who discovered the key components of second-generation antiretrovirals, the HIV/AIDS drugs used in rich countries today. To an auditorium filled with people overwhelmingly of the opinion that increasing access to drugs would require reform of intellectual property (IP) laws, Liotta described an intriguing alternative approach. With passion in his voice as he methodically laid out his argument, Liotta made a case that the problem was not IP per se, but who owned it. He argued that the fact that IP is overwhelmingly owned by those in rich countries makes access to its products unaffordable to those in poor ones. Liotta argued that if drugs were discovered in developing countries and if companies based in those countries owned the IP, the drugs would be affordable to the poor and would be relevant to their needs.[2] Building on this premise, he described two ways that he was trying to make it possible for African scientists to discover innovative drugs.

He first outlined a conventional model of knowledge transfer in which he was involved: a postdoctoral program that brought South African scientists to Emory to be trained in drug discovery.[3] Yet Liotta pointed out that if the training happened without regard to the South African economy, it would only exacerbate brain drain. The scientists' skills would be unemployable in their own country if there were no companies in South Africa that could hire drug discovery scientists. He then described an intriguing complementary approach, exemplified by another initiative in which he was involved: a company that he was in the process of launching, called iThemba Pharmaceuticals, which was going to do research on diseases of the poor in South Africa. The start-up would be set up as a for-profit company, but with lower labor costs than such a company would have in a place like the United States, and with a public mission that would tolerate the lower profit margins of treatments for diseases that disproportionately affect the world's poor. The company would be a way to build scientific knowledge capacity in Africa, mitigating brain drain while also, it was hoped, finding new drugs that would be affordable to the global poor and relevant to their needs. I was immediately intrigued by an element of this project that struck me as unique: it linked the availability of drugs with the capacity for scientific knowledge making while paying attention to place. And, as I will describe, place was simultaneously geographical, political, logistical, and imaginative.

iThemba was a small company on the outskirts of Johannesburg that opened its doors in 2009, a fresh start after an earlier Cape Town–based effort had foundered. It was cofounded by Liotta and other elite scientists to find new drugs for TB, HIV, and malaria. It received start-up funding from the South African government, which owned half of the company. iThemba's founding mission, "inexpensive therapy for infectious disease through innovative chemistry,"[4] was a distinctive one. To illuminate this distinctiveness, in this chapter I will contrast iThemba's approach with that of the three most prominent discourses of pharmaceuticals and the Global South: access to medicines; bioprospecting; and clinical trials. Putting iThemba into comparative relief with these disparate sets of discourses reveals a common element among them. All three have an implicit reliance on a problematic conceptual bifurcation between Global North and South. South Africa itself is in many ways betwixt and between Global North and South, and this small drug discovery company's mission helps to illuminate some of the limitations of pharmaceutical knowledge-making projects that take that global bifurcation for granted.

Placing Pharmaceutical Knowledge Making: Beyond Access-to-Medicines Campaigns

In Liotta's comments at the conference on human rights and the law, I was struck that iThemba's approach differed from most approaches to solving the urgent health needs of Africans, which all frame the lack of access to drugs in poor countries as a failure to meet the needs of the Other. These initiatives are widely known as access-to-medicines campaigns, using the terminology of the campaign spearheaded by the nongovernmental organization Doctors without Borders, which aims to both increase access to existing drugs in poor countries and address the lack of investment in research and development directed toward treatments for the world's poor.[5] Here, I will describe ways in which wide-ranging access-to-medicines campaigns—whether they promote distribution of generic drugs or incentivize researchers in the Global North to focus on diseases of the poor or attempt to do both—perpetuate global divisions in access to knowledge and power. Global health discourse recapitulates a troubling colonial legacy when "it configure[s] through language 'others' who would be the objects of research and the recipients of redistribution."[6] Access-to-medicines campaigns can reinforce a bifurcation of the world between (1) knowledge creators who have a moral duty to create and share knowledge and (2) those in desperate need.

This geographic bifurcation between places of knowledge production and places where the only question is access is pervasive. For example, Universities Allied for Essential Medicines emphasizes the moral emergency of the lack of access to existing drugs and lack of research into diseases of the world's poor.[7] Their solutions center on North American and European research universities sharing the fruits of their pharmaceutical knowledge with the Global South. Their motto—"our drugs, our labs, our responsibility"[8]—is at once compelling and peculiar. The "our" locates but also circumscribes, implying that the only place that knowledge can be produced, and the only place that responsibility can be located, is in the Global North.[9] The slogan reinforces the Global North's ownership claim to scientific knowledge at the same time that it advocates widening the scope of that knowledge's beneficiaries.

Social movements around generics as a solution to global health often exemplify taking the North/South bifurcation for granted, explicitly arguing for a framework in which intellectual property would be protected in rich countries but not in poor ones, so that Big Pharma could earn its profits in the former while the latter would be allowed to share the benefit.[10] Many of these initiatives build on the legacy of the Treatment Action Campaign (TAC), an important South African social movement that was highly active and visible during the late 1990s and early 2000s and that successfully won access to generic antiretrovirals imported from other developing countries in the face of opposition from global pharmaceutical companies. This social movement victory was a powerful "globalization from below" that won access to the fruits of scientific knowledge. But as TAC fought against the "dissident science" of HIV denialism that was prevalent among South African leaders at the time—in which prominent government officials all the way up to President Thabo Mbeki cited marginal scientists' skepticism that HIV was the cause of AIDS as a reason for skepticism about the efficacy of antiretrovirals—TAC organizers had good reason not to advocate the democratization of knowledge production itself.[11] The embrace of the importation of generic antiretrovirals as the solution to AIDS was a "strategic reductionism."[12] TAC's emphasis on access over research was well founded: generic drugs can often be practical solutions for addressing immediate health needs of the poor. However, these symbolic politics have contributed to giving generic drugs a virtuous sheen in global health discourse, which is misleading because they are products of for-profit companies that are constitutive of, rather than outside, global property regimes.[13] One cannot deny the real value of the drugs that companies in India and a handful of other countries supply, but this system is not the same as global democratic openness.

Generic production is itself a regulatory framework. In the global pharmaceutical production system of which generics are part, the legal ability to make particular drugs without a patent does not mean that anyone can make those drugs. As anthropologist Cori Hayden has argued, while generics provide a "counter-model to the expansion of exclusive property rights," they also create constitutive limits of "the proper copy."[14] Unbranded drugs are not the same thing as unregulated drugs, and generic producers have a stake in robust regulatory processes that allow them to play the role of trusted producers. Generic-pharmaceutical companies are important but noninnocent actors in the access-to-medicines space. Generic-pharmaceutical companies might ally with civil society activism and indeed are often vital partners for patient movements because they are well positioned to be affordable providers of the essential drugs that patient movements demand. However, generic producers can also co-opt patient activism, manipulating that political will to their own commercial ends.[15]

The framing of Indian generic-pharmaceutical manufacturers as acting in the service of the developing world might itself be understood as a successful example of generic manufacturers "unbranding" but then "rebranding" drugs[16]—rebranding the products of profit-driven Indian generic industries as virtuous charity. This rebranding draws on a sentimentalized dream of Afro-Asian solidarity in ways that obscure ongoing hierarchies rooted in colonialism that situate dependent Africans "both below Indians in civilizational terms and behind them in temporal terms."[17] Indian generic companies are symbolically and materially linked with charity, both because of their virtuous sheen and because they are the overwhelming providers of pharmaceuticals distributed by aid programs. However, charity does not always carry a positive valence from the perspective of its recipients. This is true both materially and symbolically: Indian companies disproportionately send inferior-quality drugs to Africa,[18] and like pharmaceutical consumers elsewhere, South Africans often perceive free and generic drugs as less effective.[19]

There is arguably considerable innovation involved in the reverse engineering done by these generic-pharmaceutical companies. However, even the reverse-engineering form of innovation turns out to be geographically limited, dominated by just a handful of countries (especially India and China but also Brazil and a few others).[20] South Africa now has many generic-pharmaceutical companies, but these companies are materially dependent on global trade—especially with India—because they do not have the capacity to make the active pharmaceutical ingredients locally.[21] These key components of the drugs must be imported in order for the South African companies to formulate, package, and distribute the finished drugs, and this in turn leads to considerable vulnerability to the supply chain.[22] The generics sector is now highly competitive and low margin, and it is very hard for new companies in new geographic areas to enter. Even while iThemba explored business models that included generic manufacturing, its central mission was to discover innovative drugs for which it could own the patents.

At the same time, there are inherent limitations to what generic-pharmaceutical companies can provide. Relying on generic production of first world drugs means that the developing world can get only copies rather than novel drugs and thus that the needs of the poor will not set the priorities for novel-drug discovery. This positions the global poor in a way that excludes them as potential beneficiaries of cutting-edge innovation.[23]

Generic drugs cannot fully escape their symbolic associations with the second-rate, and that matters for how they are positioned as panaceas for the poor.[24] When technologies are designed and made specifically for the world's poor—as in the case of humanitarian technologies—they are generally made for what Peter Redfield (following Fiona Terry) has called the "second-best world."[25] In this logic, it is taken as a given that the poor cannot have the ideal things, and alternative low-cost things are devised to make do. Paul Farmer's Haitian interlocutor was far more direct: "Do you know what 'appropriate technology' means? It means good things for rich people and shit for the poor."[26] Old drugs might not necessarily be lower tech than new drugs, but the same logic applies. The assumption that the first world's side projects and leftovers are good enough for the rest of the world is a problematic one.[27]

A key difference between iThemba and the movements around generics that I have discussed so far is the emphasis on research. In addition to "inexpensive therapy" and "infectious disease," there is an additional *i* phrase in iThemba's mission statement: "through innovative chemistry." The goal is not just to subsidize or lower the cost of existing drugs but also to discover new ones. In this sense, iThemba's efforts are aligned with those of the Gates Foundation, whose philanthropy supports both drug distribution and drug discovery.

The Gates Foundation's drug discovery work has overwhelmingly been channeled through product development partnerships (PDPs), a now-dominant global health drug discovery model that combines "philanthropic and state funds to support and subsidize drug discovery and development efforts, typically by commissioning research from pharmaceutical companies and academic institutions in Europe, Australia, and North America."[28] These mechanisms are of a piece with the broader financialization of global health, in which philanthropic and state funds are seen as investments not only in health broadly but in intellectual and material products with a potential for return on investment.[29]

iThemba participated in these PDP initiatives in a small way, albeit not as a "partner" but as a contract research organization. Contract research organizations that work in later-stage drug development, carrying out clinical trials for pharmaceutical companies in a fee-for-service model, are prominently discussed in the anthropology of pharmaceuticals in the Global South.[30] Such organizations in the earlier-stage drug discovery space are less widely analyzed. These early-stage contract research organizations are chemistry-based companies that synthesize or analyze particular molecules for a fee. iThemba did this type of contract research as a way to earn money to sustain itself and support its in-house drug discovery initiatives. Its contract research clients included some of the most prominent PDP efforts, including the Doctors without Borders–initiated Drugs for Neglected Diseases Initiative and the Gates Foundation–dominated Medicines for Malaria Venture.[31] Yet contract research does not really count as partnership: contract research organizations carry out much of the work of these projects, but they are rarely considered partners, because they simply provide service for a fee and do not participate in setting the research agenda.[32] Thus, they maintain rather than trouble global hierarchies.

iThemba's peripheral status in these PDP efforts signals a deeper important element of difference. Gates and others have paid scant attention to *who* is making the science, and his organization's funding has preserved the Global North as the privileged location for innovative knowledge making. Even as they occasionally brandish photographs of African scientists (albeit much more rarely than they brandish photographs of African patients), his and other philanthropists' preference for funding scientists with "track records" provides a tremendous advantage to scientists in the Global North, and the Gates Foundation has prominently partnered with big pharmaceutical companies for neglected-disease research.[33]

In the system of global flows that Gates has participated in creating, geography is central to organizing technological products: the system is designed such that cheap drugs and cheap computers flow to the Global South without disrupting high prices in the Global North. But the geography of technological innovators has little role in Gates's model. The research could conceivably be done anywhere, and it's easier and less risky to simply keep it where it is. From the Gates Foundation's perspective, present trends in where funded researchers are located might as well continue.

The distribution of funds through Gates Foundation programs such as the Global Grand Challenges, "a family of initiatives fostering innovation to solve key global health and development problems" in which pharmaceutical research is prominent, is telling.[34] The overwhelming majority of grants given through the Grand Challenges initiative have been given to scientists working in North America: in a report dated December 2012, the Grants Map on the Gates Foundation website listed the countries receiving grants as follows: North America, 662; Europe, 149; Asia, 75; Oceania, 48; Africa, 36; South America, 16.[35] In light of the mismatch between the countries where global health and development problems are most urgent and those where research is being supported, the Gates Foundation's grants exemplify the presumption of the unidirectionality of global knowledge flows.

Drug discovery becomes an exemplar of the broader phenomenon that historian Noémi Tousignant has articulated: "One of the ways in which colonial policies, geographical imaginations, global economic inequality and intellectual property rights have helped to create and maintain peripheries is by reserving certain kinds of scientific practice for locations in which wealth and power [are] concentrated."[36] Even if the goals of drug discovery were to be set with the needs of resource-poor populations in mind, the fact that drug discovery is concentrated in rich countries perpetuates center/periphery bifurcations.

Philanthropic models for addressing social needs often preserve or exacerbate inequality rather than mitigating it. As Linsey McGoey points out, "Philanthropists themselves are often the first to admit that their philanthropy is aimed at preserving rather than redistributing wealth. Carlos Slim, a Mexican business magnate who is among the world's richest men,[37] is perhaps the most candid about this fact, summarizing his own approach to charity with the comment, 'Wealth is like an orchard. You have to share the fruit, not the trees.'"[38] Rather than raising the specter of the redistribution of land, maintaining a monopoly on the generation of pharmaceutical knowledge allows those in rich countries—even well-meaning university students—to occupy the role of the orchard owner. But unlike land, knowledge-making capacity is not an inherently limited resource, and those who support access to medicines should welcome the spread of both fruit and trees.

One objection to South African investment in pharmaceutical R&D that is often raised, both with interlocutors in South Africa and with audiences in the United States and other wealthy countries, is that since many South Africans still lack access to existing drugs, it is more efficient to focus on provision than on new-drug development. The crisis of HIV/AIDS, so pervasive in South Africa, features prominently in this kind of trade-off argument: people are dying today, and untreated people are spreading the disease today, so there is no time, let alone resources, to invest in possible future treatments.

This argument for prioritizing present treatment over research for future treatment has a certain appeal, but it is also troubling. First, it is a naïve way of framing decisions about investment: the notion of a single pot of money that sits somewhere waiting to be allocated to either science or health care may be true for the Gates Foundation, but it's not true for governments and other payers. In the absence of investment in pharmaceutical R&D, there is no reason to assume that those funds would otherwise go to basic health care provision. Money not spent for pharmaceutical R&D might be spent instead on completely different spheres. For example, the South African government has invested significant sums in astronomy research (the Square Kilometre Array radio telescope), on the one hand,[39] and information technology development, on the other. South African pharmaceutical R&D investment in diseases that have a high impact in South Africa is small not only in absolute terms but also as a share of R&D spending overall—and arguments for changing that agenda draw on reasoning based both on social justice and on the potential long-term return on investment.[40]

One might conceivably argue that further research is not necessary anyway, since HIV, TB, and malaria all have existing treatments. Yet existing treatments for malaria are inadequate, those for TB are onerous and thus prone to cultivating resistance, and there will eventually need to be a new generation of HIV treatments. It should not be assumed in advance that future innovative drugs for these diseases will emerge exclusively from the companies in the same geographic boundaries that produced past treatments.

Second, the objection that countries with patients in need of treatment today should invest in that treatment rather than in research also exemplifies a patrician Orientalist frame that holds poor countries to purer moral standards than rich ones.[41] If expensive R&D can be ethically pursued by a country only once basic needs are met, that would rule out R&D in the United States. Notwithstanding its high per capita wealth, the United States also has populations who struggle with poverty and face barriers to access to existing drugs, and yet that is not a justification for forgoing research.

What unites all these philanthropic and activist movements for global access to drugs is that their critique has overwhelmingly been framed in terms of failure to meet the needs of the Other: scientists and drug companies in the Global North failing to rise to the humanitarian challenge of meeting the needs of the Global South. This assumes an immutability of the nature of the global flows: knowledge will be made in the Global North, and the challenge is to direct that research toward better priorities and to spread the benefit to the Global South. But what if we don't take the unidirectionality of knowledge flows for granted?

Placing Pharmaceutical Knowledge Making: Beyond Bioprospecting

In addition to the global health literature in which access to medicines is a central concern, a second major discussion of pharmaceutical knowledge with regard to the Global South involves traditional knowledge, botanical products, and bioprospecting,[42] intertwined topics that have a long history rooted in colonial relations.[43] The power dynamics of bioprospecting in

South Africa in particular are shaped by the country's settler-colonial roots and legacy: white and Asian scientists working in governmental labs have long taken black botanical knowledge as a field of study.[44] In this historically racialized hierarchy, bioprospecting in South Africa can be seen as a microcosm of global inequalities, even as bioprospecting becomes a site of hope in the nation-building projects of the "African Renaissance."[45]

Bioprospecting emerges as a prominent site of inquiry in postcolonial science and technology studies (STS) because that field emphasizes paying attention to indigenous knowledges in the context of colonial and postcolonial relations. For example, bioprospecting and traditional knowledges are amply represented in the field-defining collection *The Postcolonial Science and Technology Studies Reader*.[46] This focus does important work, but it defines the scope of the potential for drug discovery in the Global South too narrowly. Most drug discovery globally does not draw on botanical products or traditional knowledges. The approach has always been a very small portion of research and has become increasingly marginal: as National Institutes of Health biodiversity researcher Joshua Rosenthal lamented in his review of Cori Hayden's important ethnography of bioprospecting in Mexico, "it seems that we might have already reached the point where there are more people analysing ethnobotanists and pharmacognosists than there are practitioners in these areas."[47] Here I explore how analysis of Global South drug discovery efforts based on mainstream synthetic chemistry can help to expand the frame and unsettle North/South bifurcations.

Anthropological and STS accounts of the intersections of traditional knowledges and pharmaceutical science reveal how traditional knowledges and biomedical ones are mutually constituted, and yet one of the things that this literature brings to the fore is the conflict between *local knowledge* and *global science*. This conflict is not simple. On the one hand, traditional African botanical knowledge was never as local as the discourses of "indigenous knowledge" and "bioprospecting" have implied, since both plants and knowledges have traveled around Africa long before and well beyond varied pharmaceutical companies' and government research initiatives.[48] Furthermore, traditional knowledges and biomedical knowledges in African places inform and shape one another, as traditional healers draw on pharmaceutical knowledge and vice versa, even as the epistemologies of traditional healing and pharmaceutical research remain incommensurate.[49] As anthropologist Damien Droney has observed with regard to capacity-building projects around herbal medicine in Ghana, "The science of herbal medicine does not simply increase knowledge about an unambiguous object. It shapes herbal medicine as it studies it."[50] Along a related line of analysis in Kenya, anthropologists Wenzel Geissler and Ruth Prince track the unsuccessful attempts of ethnobotanists to "search for universals in atomic specificity" that could "do away with relation and process."[51] Divisions between scientific knowledge and indigenous knowledge are not as neat as they might seem.

Yet there is a fundamental tension in anthropological and STS accounts of bioprospecting between biomedical scientists (including African scientists) and traditional healers. Both cooperation and conflict between traditional healers and biomedical scientists occur within a hierarchy: insofar as traditional knowledge of plants constitutes Global South knowledge, it is only local knowledge rather than science, unless it is brought into a scientific rubric. In the case of pharmaceuticals, a key step is transforming "raw materials" into "active ingredients" through material chemical processes in laboratories. This transformation allows botanical products to become the basis of pharmaceutical knowledge and, relatedly, to become globally mobile pharmaceuticals.

Bioprospecting and ethnoscience approaches exemplify one type of translation model: translating local knowledge into global science. Whereas ethnographic exploration of efforts to pharmaceutically exploit botanical products and traditional knowledge foregrounds the conflict between local knowledge and global science,[52] the iThemba project sidesteps this binary. iThemba offers an opportunity to explore a Global South site that sought, not to translate local knowledge into global science, but to participate in global science in the same way that northern scientists do.

Importantly, the knowledge that iThemba sought to create was not based on traditional knowledge or plant-based therapies. When I have spoken about my research on South African drug discovery, audiences in rich countries routinely either assume that iThemba did bioprospecting or suggest looking at plants and traditional knowledge for ways to turn those into medicines. These were not iThemba's approach. Rather, its methods and materials were drawn from mainstream synthetic chemistry. Of course, a promising plant-based lead molecule would not have been rejected. Indeed, iThemba's chief commercialization officer, David Walwyn, had previously had a company that sought to find novel lead compounds against HIV by using chemistry methods to test the efficacy of the botanical products that traditional healers used in their treatment regimens. The products were originally brought to the government-run Council for Scientific and Industrial Research by the healers themselves, who wanted scientific verification of their claims.

As Walwyn told me in an interview in 2010, that company, Arvir, was a failure, for reasons that had nothing to do with IP: although the plant compounds had some efficacy against HIV, none had potent enough efficacy to be made into a pharmaceutical. iThemba acquired Arvir's technologies and its last two employees in 2010, and the employees moved on within a couple of years. Other iThemba scientists also raised the prospect of botanical products as a resource in interviews. A few of the scientists mentioned interest in the model of hoodia, a plant traditionally used by the San people and the origin of a compound patented by South African government researchers as a base for a potential appetite suppressant drug;[53] notably, the iThemba bench scientists did not seem aware of the controversies around biopiracy in that case.[54] However, local plants and traditional knowledge were probably no more central at iThemba than they are in the pharmaceutical industry more broadly.

iThemba's notion of the value of African research was thus discontinuous with what historian Helen Tilley has described as the early twentieth-century imperial sense of Africa as a "living laboratory."[55] In Tilley's account, doing science in Africa was argued to have particular value because of specific and intimate access to the phenomena of nature there. The "living" quality of the laboratory in Tilley's account comes from the field sites of science rather than being grounded in laboratory-based scientists and their practices. In contrast, in iThemba's model, Africa was envisioned as a laboratory in the far more mundane sense that laboratory scientists were living and working there.

We should be suspicious of the appeal that ethnoscience holds for STS of pharmaceuticals in the Global South. Ethnoscience emerged out of 1960s anthropology as a way to refer to the anthropological study of human knowledge systems from the perspective of those being studied and is especially attentive to indigenous modes of classification of kinship and natural phenomena.[56] A parallel literature within STS has been interested in the rationalities of domains of knowledge generally not considered scientific, such as indigenous navigation of the Pacific Ocean, to emphasize the heterogeneity of both traditional knowledges and of science.[57] It is valuable to pay attention to what counts as systematic knowledge in particular cultures, and this is part of what is accomplished in the study of traditional knowledges. Yet as worthy of study as bioprospecting and traditional knowledges are, their centrality in the field is of a piece with romantic visions of global technoscience's Others.[58]

Operating as if *the* problem of southern knowledge making is managing traditional knowledge and natural resources is symptomatic of the epistemological bifurcation that has pervaded analysis of IP and medicines in the Global South. In global health discourse, the assumption has been that global knowledge will be made in the North and shared with the South (or not). In traditional-medicines discourse, the assumption has been that local knowledges already exist in the South and will be exploited by the North (or not). Neither recognizes the capacity of the Global South to participate in the creation of new global knowledge. Analysis of iThemba thus also complements new work on genomics in the Global South.[59] To an even greater degree than in that genomics research, which relies on locally derived biological materials, the Global South knowledge making at iThemba was place specific without being autochthonous.

Placing Pharmaceutical Knowledge Making: Beyond Clinical Trial Research

In addition to access to medicines and bioprospecting, the third well-analyzed area of pharmaceuticals and the Global South is clinical trials.[60] It is another area with which most audiences I speak with about my ethnographic work on South African drug discovery are familiar. In academic literature, there is a large body of critique that posits "therapeutic domination" as an exemplar of contemporary neocolonial power.[61] More broadly, both academic and popular audiences in rich countries often respond to the topic of pharmaceuticals in Africa by lamenting the abuses of disempowered African research subjects by Western researchers. This is a misapprehension of the main ethical problems with clinical trials in Africa. As anthropologist Wenzel Geissler points out, "Contrary to dark tales of the *Constant Gardener* type . . . in which Euro-American scientists, paid by inscrutable global corporations or, with equally sinister connotations, the U.S. government and army, experiment on black people to generate profits from vulnerable, disposable bodies, most medical research in Africa today is open to public scrutiny, publicly funded (albeit not necessarily by the national government of the population enrolled in the trial), and by and large free from immediate corporate profit interests."[62] Yet if shadowy conspiracies are not a particularly pressing issue in the landscape of clinical trials in Africa, there are many other concerns. Here, I will focus on two, which help to show how iThemba's project is distinct: the reliance of clinical trials on inequalities and the epistemological status of clinical trials as secondary to drug discovery.

In clinical trials in the Global South, it is generally the claims of scientists in the Global North that are being tested on bodies in the Global South. Transnational clinical trials in Africa rely on inequalities of both patients and researchers, and African researchers are highly aware of the "normal inequalities" that structure their collaborations: "local standards" dictate lower remuneration for African collaborators, but at the same time, it is the rich-country collaborators who get to decide how and when global standards must be met.[63] As with bioprospecting research, these global inequalities can play out in microcosm in South Africa, because it has a much more highly developed economic and regulatory infrastructure than other African countries, and thus a higher capacity to do clinical trial research, but still has millions of people with limited access to medicine and who thus make particularly appealing trial subjects because they have not previously received treatment for their disease.[64]

Insofar as African physician-researchers are participating in making knowledge through their collaboration in clinical trials, it is overwhelmingly *not* at the research design stage. This state of affairs reflects the broader postcolonial discourse in which Africa is not fully part of the human project but rather "an object of experimentation."[65] The capacity of African research subjects to create their own knowledge is doubly erased,[66] as even local trialists have limited ability to make truth claims. Investigators in the Global North might debate the ethics of that experimentation among themselves, but they do so in ways that leave global imbalances of both epistemological power and control over resources intact.[67]

Transnational collaborations often follow a "core-periphery model," in which the African researchers are "considered as the 'field workers' who collect data following the study protocol design and implementation by Western partners."[68] The priorities and specifically located perspectives of the African investigators in the partnership are ignored as the research gets put to the service of globally mobile global health discourse.[69] Although there can be social value in transnational research collaborations for researchers, research subjects, and even research and health infrastructures,[70] African collaborators are not recognized as the true global knowledge makers. Frequently, data are extracted from the South and analyzed in the North, which fuels northern science and exacerbates North/South disparities in research capacity rather than ameliorating them.[71]

Similarly, African research subjects are not imagined to be true objects of global knowledge. Sometimes, like trial subjects in India, African research subjects are stand-ins for Global North consumers and patients, not configured as future consumers and patients themselves.[72] Other times, the studies are meant to reveal whether "already established treatments" might work in "resource-poor settings."[73] In either case, the inequalities studied by global health become, as anthropologist Johanna Crane points out, *valuable inequalities*: "Within global health, the very characteristics that once led some Western experts to dismiss HIV treatment in Africa as unwise—impoverished patients, poor infrastructure, understaffed health facilities—are now those that make many African countries attractive 'resource-poor settings' that can offer 'global' research and educational opportunities unavailable in 'resource-rich settings' like the United States."[74]

African clinician researchers have much to gain as intermediaries in these relationships between global health studies and local patients, but their role is highly circumscribed. Clinical trials in Africa are rarely designed or led by Africans—even in South Africa, with its highly developed scientific sphere relative to the rest of the continent. As such, clinical trials are less meaningful as a locus of African knowledge making than drug discovery would be.

This leads to a second epistemological issue at stake that makes iThemba's project distinct from clinical trial research: original drug discovery is more scientifically fundamental than clinical trial knowledge and, as such, makes a stronger claim to pharmaceutical knowledge making.

The epistemological status of clinical trial knowledge has an ambivalent character. On the one hand, randomized controlled trials have become "the gold standard" of pharmaceutical knowledge making.[75] On the other hand, such trials simply test the products and claims of earlier, more fundamental stages of pharmaceutical research—conceptually, they follow from, rather than constitute, "drug discovery."

Either way, in clinical trial research, African researchers don't really have access to the full credit: their work in clinical trials is unlikely to be accepted as "gold standard" research unless it is under the supervision of transnational researchers, and they rarely get to participate in the more fundamental stages of drug discovery. To put the issue more conceptually, in transnational clinical trials generally, people in the Global South are *objects* of science (the ones about whom knowledge is made) rather than its *subjects* (the ones making knowledge).[76]

The iThemba story is not about clinical trials or the power differentials and ethical problems that emerge in late-stage pharmaceutical research. The scientists at iThemba sought to participate in the creation of original knowledge claims at the very beginning of the drug discovery value chain, at its most scientific level. They wanted to be at the beginning of pharmaceutical knowledge making, in both time and space: forming the initial hypotheses about novel compounds, synthesizing them, researching them preclinically, and creating rather than implementing downstream study. The knowledge created would not be merely practical technical knowledge that would test or elaborate a drug efficacy claim. Rather, it would be conceptually driven technoscientific knowledge at its most basic. It would be able to go out into the world in the same way that knowledge created by any other drug discovery enterprise would, whether at GlaxoSmithKline or at Emory University. This presents very different epistemological issues from the kinds raised in exclusive focus on late-stage clinical trials. The creation of such knowledge would instantiate the hope of being a fully equal collaborator in global projects of drug discovery.

Conclusion: Beyond Diffusion and Translation

iThemba provides an opportunity to consider African knowledge making beyond diffusion and translation.[77] Following sociologist Logan D. A. Williams, I argue that iThemba is a site that provides an important opportunity to foreground knowledge circulation within the Global South, without seeing that knowledge as derivative of a northern original. The case of iThemba offers a chance to consider other registers of what it might mean to study technoscience in the Global South beyond diffusion and translation. Diffusion is a key aspect of access to medicines, translation is a key aspect of traditional knowledge and bioprospecting, and both are important in clinical trials in Africa. Beginning an analysis with iThemba provides an opportunity to start elsewhere.

The three discursive spheres that I have discussed in this chapter—access-to-medicines campaigns, bioprospecting, and transnational clinical trials—each bring up distinct ethical and epistemological questions. Yet what unites them is a bifurcation of the world between North and South, in which Global South researchers are not in a position to make the kinds of universal scientific claims that Global North researchers can. Problem definition is the purview of the North, in each case. And the traffic flows in only particular ways: the North is the source of scientific questions and final solutions, while the South is the source of raw materials and subjects on which those can be tested.

Troubling these bifurcations connects with larger projects of postcolonial technoscience, as historian Warwick Anderson has articulated: "A postcolonial analysis thus offers us a chance of disconcerting conventional accounts of so-called 'global' technoscience, revealing and complicating the durable dichotomies, produced under colonial regimes, which underpin many of its practices and hegemonic claims."[78] Many of the binaries that he spells out are pervasive in the global health discourses: "These binaries still operate in terms of global/local, first-world/third-world, Western/

Indigenous, modern/traditional, developed/underdeveloped, big-science/small-science, nuclear/non-nuclear, and even theory/practice."[79] Analysis of iThemba provides an opportunity to explore a Global South science site that resists placement on either side of the binaries.

We might return to Dennis Liotta's intervention at the human rights and the law conference: that the problem is not necessarily intellectual property per se, but that the Global North has a monopoly on the creation of intellectual property. Even as there is much that this framing leaves out, it points to both an important problem and an intriguing site of hope: we can acknowledge the historic and present concentration of pharmaceutical knowledge making without accepting that as an eternal state of affairs.

TWO

In the Shadows of the Dynamite Factory

The physical site of iThemba Pharmaceuticals provides a useful entrée into the company's postcolonial situatedness. The company was located in a Johannesburg suburb called Modderfontein. As of the early twenty-first century, the part of Modderfontein in which iThemba was located was a tranquil suburban office park, with narrow streets and eucalyptus trees.[1] Yet the heavy industrial past of the site is visible as soon as one starts to look. The whole suburban office park is at the edge of a huge lot that is a complex of African Explosives and Chemical Industries (AECI), where explosives for the mining industry are still produced and tested. It was the site of a historic Nobel dynamite factory (fig. 4)—the largest in the world in its early twentieth-century heyday. A couple of minutes' walk from iThemba's building, the AECI still maintains a dynamite museum, which Johannesburg Tourism promotes as a tourist site,[2] in the house of an early executive of the dynamite company.

The dynamite company casts two sets of shadows: spatial and conceptual. First, the dynamite factory provides a way to locate iThemba in literal geographical space. Second, dynamite becomes a way to locate iThemba's endeavor as emerging out of fraught histories that both constrain and create possibilities for postcolonial science: colonial legacies of extraction that have contributed to the global pharmaceutical industry, the South African economy, and more. Thus, the legacy of dynamite in South Africa provides a useful vantage point from which to examine the contemporary history of pharmaceuticals there. At the same time, the location of iThemba Pharmaceuticals on land with an explosive legacy provides a route into consideration of the history of the pharmaceutical industry more generally and the contours of South Africa's broader, explosive-dependent history and economy.

4. Early dynamite box emblem from the Alfred Nobel Modderfontein Factory, on display at the Dynamite Company Museum. Photograph by Katherine Behar.

The dynamite factory was set up well outside the city because of the risk of explosions. A wildlife park provided a buffer between it and the city center, hundreds of acres of which remain a nature reserve.[3] Along the way, many manufacturing plants joined the dynamite factory, and today many parts of Modderfontein are heavily industrial. There was and is a large township adjacent to Modderfontein, Alexandra, which is situated between Modderfontein and what has become the important, prosperous suburb of Sandton, an area that has now superseded the historical downtown of Johannesburg as a site for shopping malls, hotels, businesses, and residential complexes. Now the sprawl of Johannesburg is increasingly closing in on the site from the south and filling in the surrounding areas as well, predominantly with gated communities of single-family homes that are quickly covering the hillsides.[4]

That iThemba exists in the shadow of this dynamite factory is simultaneously a contingent logistical fact, a clue to the historical links between the global pharmaceutical industry and the mining and munitions industries, and a window into the situatedness of drug discovery in South Africa. In this chapter, I will first describe the contingent history of the literal placement of iThemba in Modderfontein and the reflections of one of the bench scientists on its location. Then, I will work through multiple ways of understanding the virtual shadows, meditating upon the resonances between dynamite sticks and pharmaceutical tablets; the route to capitalism that dynamite built; parallels between dynamite and antiretroviral manufacture; and contemporary mining as a context for iThemba's work. In dynamite's infrastructural and symbolic shadows, there are ambivalences between destruction and productivity that pharmaceutical scientists hope to redeem.

iThemba's Place in an Explosive History

There are contingent, logistical reasons that iThemba's labs were on AECI land. One of the South African cofounders of iThemba and former chairman of its board, Frank Fisher, had been a longtime group research manager and project director for AECI. His own work had been in areas unrelated to drug discovery (paper and plastics), but as research director of the whole AECI group he had set up a biotechnology function at the Modderfontein site.[5] Fisher had come to know British scientist Tony Barrett, who would become another cofounder of iThemba, because AECI had employed him as a consultant over the years. It was through Barrett's networks that Fisher would meet the other scientists who would become cofounders of iThemba. Years later, when cofounder Dennis Liotta asked Fisher to secure lab space in Johannesburg for iThemba, Fisher chose the Modderfontein labs that he had helped to set up for AECI.

At the same time, that the company exists on AECI land is a clue to the situatedness of drug discovery in South Africa: in the shadow of the mining industry. In some sense, this explosive history is merely part of the background scenery of the site. Yet, in contrast to the likely situation in a place where history is less a topic of broad public engagement (such as the United States), the scientists at iThemba were all aware of this history, and some were quite interested in it. This is of a piece with South Africans' generally high awareness of and engagement with history, which might be related to being citizens of a young country whose history has been (re)written in their lifetimes. In an interview in 2010, one bench scientist at iThemba described the site's history in a way that was complex and multilayered enough to justify quoting here at length.

The scientist began to lay out the history by highlighting the scope of mining and of dynamite in South Africa: "Just to give you an idea of the history of this area by the way. Alfred Nobel, when he patented his dynamite manufacture, it was made just behind us. And we made a lot of it. The mining industry was just booming here." Yet the scientist's tone shifted immediately, when noting mining's exceptionalism and lamenting that pharmaceuticals have not followed in the footsteps of that boom: "And that's the problem that we've had in South Africa, that we have really just been a formulation economy apart from mining. And in the pharmaceutical industry, this is now 120 years later, we are still a formulation economy in terms of pharmaceuticals. We need to move away from that and actually start making things." The valorization of "actually making things" is common enough in both masculinist and nationalist discourses of manufacturing, and it is doing interesting work here as a foil for merely formulating things (i.e., assembling them). Unlike "extraction economy" and "knowledge economy," "formulation economy" is not a widely used term. It's essentially a neologism to describe an unsatisfying manufacturing economy that assembles pre-made parts into finished goods rather than building a whole product.

The scientist returned to the mining case, to describe the mechanism by which it provided the impetus for South Africans to stop being dependent on imported materials and to start to make many different things:

> So now I think that dynamite, to give you an idea, we developed a lot of industries around it. And it was all started by mining companies because they needed dynamite. And the problem was it took too long by ship; there were too many wars going on, in 1900, so they needed a source of glycerin. And at the time there was a company in Cape Town making soap, so they bought the glycerin there. And then they said, well, paper, we are using a lot of paper for the dynamite, so we need to make our own paper, so Sappi started up—South African Pulp and Paper Industry. So the mining industry basically bought it; if they needed anything, they just bought it. We've done a bit of fine chemicals along the way. Especially Sasol with their coal-to-fuel provision, and even that was quite controversial: they bought it from Nazi Germany, that knowledge. But they took that and they developed it.

In this telling, mining spins off into success for both private industry (Sappi) and state corporations (Sasol). The scientist lamented that the pharmaceutical industry in South Africa was still largely stuck in the formulation stage that characterized the earliest era of the mining industry, rather than following its predecessor's path onward to manufacture and innovation: "But again the pharmaceutical industry, nothing happened. So it's just interesting that we were [starting] a mining industry, doing a lot of formulation; we just imported everything, and then mixed them together and sold them off. And the pharmaceutical industry is still doing that. So that's why it's exciting what we are doing. We are trying to do something and move away from that. So it's all very interesting." This narrative of history evokes the literal and conceptual shadows of mining for would-be pharmaceutical scientists and is suggestive of both a practical and an idealized model for a potential future of the pharmaceutical industry in South Africa.

The scientist also pointed out that the infrastructure of dynamite contributes more than old buildings and an interesting historical counterpoint. Because of the ongoing manufacture of ammonium nitrate nearby—a key component of both fertilizers and explosives that can itself be highly explosive and must be handled with care—iThemba's office park never loses power: "They used to produce ammonia, a kilometer up the road. But I think the Chinese bought that in the 1970s, so we get our ammonia from Sasol. But they still make ammonium nitrate over there. So this area is a national asset, so with all the power going off, the power never goes off in this area. There's an ammonium nitrate plant going twenty-four hours a day, and you can't just stop that." That electricity can be taken for granted in this space is itself quite notable. Access to cheap electricity is a key competitive advantage of South African industries broadly,[6] and for industries such as pharmaceuticals, reliability may be more fundamental. The reliable availability of electricity is a distinct advantage for would-be South African drug discoverers, relative to those in other parts of Africa and the developing world, where they might have to generate their own electricity in order to be able to count on it.[7]

In this iThemba scientist's layered account of the history of the site in which they were working, the legacy of dynamite has left iThemba with many things: a model of a world-leading industry built in South Africa, the growth of complementary industries and a robust infrastructure to serve it, and the as-yet-unrealized promise to build a South African pharmaceutical industry that could be as successful as its mining-industry forebear. Historical ethnographer Marissa Mika has pointed out that "scholars are increasingly turning their attention to the debris, ruins, residue, and detritus of medical collaborations and experiments as a way to open up important histories of the practices, ethics, and legacies of biomedicine in colonial and postcolonial African contexts and beyond,"[8] and the story of iThemba's physical site provides a chance to reflect on the ways in which infrastructures that were not conceived of as medical can do things along the same lines.

The history of dynamite, in this scientist's rendering, is simultaneously material and symbolic. It is "imperial debris," in anthropologist Ann Laura Stoler's terms: "compounded layers" of "empire's ruins contour and carve through the psychic and material space in which people live."[9] Stepping back from the literal site on which iThemba's labs and offices were located,

I will now turn to more lateral ways in which dynamite's virtual shadows can provide insight into the complicated legacies on which iThemba hoped to build.

Dynamite Sticks and Pharmaceutical Tablets

By-products of the mining industry have been extremely important in the pharmaceutical industry not just in South Africa but globally. Writing in the second half of the nineteenth century in volume 3 of *Capital*, Karl Marx commented on the remarkable ability of the chemical industry to make use of waste: "The most striking example of utilising waste is furnished by the chemical industry. It utilises not only its own waste, for which it finds new uses, but also that of many other industries. For instance, it converts the formerly almost useless gas-tar into aniline dyes, alizarin, and, more recently, even into drugs."[10] Coal tar, as gas-tar is more commonly known, would become the source of quinoline and, later, acetaminophen (paracetamol)—to this day, the most popular pain reliever in the world.[11] The first industry to develop industrial research laboratories was the coal

5. Chemical equipment on display at the Dynamite Factory Museum. Photograph by Katherine Behar.

tar dyestuffs industry, and those would become the model for pharmaceutical research labs.[12] The deepening integration of pharmacy and synthetic organic chemistry would become the basis of the modern pharmaceutical industry at the end of the nineteenth century and into the twentieth.[13] The material legacies and intellectual legacies shared by pharmaceutical companies and other important industries of the period are intertwined.

The technology of pharmaceutical tablets and that of explosives are related to each other in plural ways (fig. 5). During my fieldwork at iThemba, this was often demonstrated to me in lab meetings in which the scientists would compare potential reagents with which to synthesize desired molecules. One of the factors that often came up was whether particular components were explosive, which would add considerably to the shipping costs. This underscored the fact that although finished pharmaceuticals are generally inert in transit, the components with which they are made can be highly volatile.

Pills are perhaps more directly similar to bullets than to dynamite sticks: the mid-nineteenth-century innovation of a punch and die system used to create pharmaceutical tablets was inspired by that used to make bullets, as exemplified by the landmark 1843 patent by William Brockedon for "Shaping Pills, Lozenges, and Black Lead by Pressure in Dies."[14] This drew on and contributed to contemporary advances in bullet-making technologies and applied those to pharmaceuticals and pencils. In other patents, Brockedon put the then-exotic material of rubber to use in firearms.[15] Advances in munitions went hand in hand with advances in other tools of industrial and colonial expansion.

In both industrially produced pharmaceuticals and dynamite, the presence of binders makes them less vulnerable to the environment and thus more stable in transportation and storage.[16] Almost all active pharmaceutical ingredients need the addition of other substances before they can be turned into mass-manufactured tablets.[17] Analogously, nitroglycerin might be understood to be an "active ingredient" in dynamite. Before the invention of dynamite, the most common explosives in use were gunpowder and nitroglycerin. Gunpowder was suboptimal for mining for many reasons, most notably because it was more effective at propelling things than at blowing them up. Nitroglycerin is a very powerful explosive, but it is unstable and highly vulnerable to accidental explosion by impact. Alfred Nobel was inspired by Ascanio Sobrero's invention of nitroglycerin, in 1847, and experimented with the compound.[18] Nobel's brother Emil died in a nitroglycerin explosion in 1864, which was among many accidents that spurred Nobel to create a safer compound, adding stabilizers to create dynamite.[19] The dynamite's stabilizers and wrapping paper might be understood to correspond to the stabilizers and layers of packaging of pills.

Nitroglycerin itself is also an active ingredient in pharmaceuticals. There is a direct connection between the two: "the vasodilating properties of nitroglycerin were discovered by William Murrell after its invention by Alfred Nobel as the active constituent of dynamite."[20] The pharmaceutical properties of nitroglycerin were recognized in part because of anecdotal observations about the effects of the compound on workers in dynamite factories—causing flushing and vascular headaches on return to work each week, until they had built up tolerance.[21] These symptoms were caused by the nitroglycerin that workers absorbed through their skin, lowering their blood pressure.[22] The effects seemed similar to those of amyl nitrate, a vasodepressor known to reduce anginal pain, and nitroglycerin could be used sublingually as a prophylactic for angina.[23]

The coincidence of the same product being used as an explosive and as a pharmaceutical could be unappealing—most notably for Alfred Nobel himself: "Paradoxically, Nobel, who suffered from angina, rejected the prescription, saying, 'Isn't it the irony of fate that I have been prescribed nitroglycerin, to be taken internally! They call it Trinitrin, so as not to scare the chemist and the public.'"[24] For Nobel, the destructive capacity of dynamite clung somehow to the pill—the physiological effects could not be completely divorced from the larger effects of the compound in the world. Nobel soon died of his cardiovascular condition.

It is also notable that the historic dynamite factory at iThemba's site was a *Nobel* factory. This means that when it was set up, it was part of the international network licensed by Alfred Nobel. Nobel's legacy is itself something of an evocative paradox. Today, his name is heavily associated with prizes recognizing achievement in science, literature, and peace and so represents hope in progress. But he made his fortune in explosives stable enough to be transported, which, while certainly a technological advance, facilitated wars and colonialism. Dynamite was patented by Alfred Nobel in 1867 and was known as "safety powder" because it was more manageable than previous powerful explosives. At the same time, dynamite was seen as so destructive a force that it earned the epithet "a most damnable invention."[25] An obituary for Alfred Nobel that was erroneously published in 1888 while he was still alive referred to him as "the merchant of death."[26] This spurred his efforts to transform his legacy, through bequeathing his wealth after his actual death (in 1896) to the Nobel Prizes.

Like Nobel's own legacy, dynamite retains an ambivalence between production and destruction. Nitroglycerin both as a pharmaceutical and as an explosive can be understood as *pharmakon*. The word "pharmaceutical" comes from the Greek *pharmakon*, a polyvalent word for remedy, poison, and a means of producing something.[27] Whereas pills act on the body, dynamite acts on the landscape, and risks and harm beyond the desired effects are inevitable. As anthropologist Emily Martin has pointed out, "side effects" and "collateral damage" are linked.[28] However, dynamite is more inherently destructive, intentionally and by design, as opposed to incidentally so. A few pharmaceuticals operate on a somewhat analogous logic to dynamite, such as cancer chemotherapy—also referred to as "cytotoxic chemotherapy," or cell-poisoning treatment. Chemotherapy's roots are in mustard gas as a chemical weapon, and so its inherent violence makes sense.[29] Most pharmaceuticals, however, aim to reorder the body more peacefully—even as they never fully avoid the simultaneity of remedy and harm.

The essential work of dynamite is to demolish something (earth or rock) in order to access something else (the desired minerals). Moreover, the fundamental nature of mining is exploitative: it exploits both the land and labor. The hope in harnessing synthetic chemistry for pharmaceuticals in place of explosives is connected to a desire to use technoscience to not only make a profit but also foster human well-being. In a country and a region so associated with destruction of peoples for profit, pharmaceuticals are a potential site of reparation.

A Route to Global Capitalism Built by Dynamite

The mining industry is central to the historical and contemporary context of chemistry in South Africa and, in some senses, to its social history as well.[30] South Africa's economy was and remains to an important degree reliant on extractive industries: those that draw raw materials from the earth. In South Africa, that meant predominantly diamonds and gold in an earlier era, and today the country is a leading global source of metals such as platinum, chrome, and iron ore, among others, as well as an important source of coal. Indeed, economic analysts have described the strengths and challenges of both the apartheid-era economy and the post-apartheid economy in terms of the overwhelming predominance of the "Minerals-Energy Complex (MEC) that lies at the core of the South African economy, not only by virtue of its weight in economic activity but also through its determining role throughout the rest of the economy."[31]

Dynamite was a foundational material for South Africa's colonial economy of extraction, and attention to it can also inform understandings of materials of a postcolonial knowledge economy—in this case, pharmaceuticals. The history of dynamite in South Africa can provide a window into contemporary hope for pharmaceuticals there, both conceptually and materially. Mining was what brought the place that would become South Africa deep into global capitalism. As anthropologists Jean Comaroff and John Comaroff point out, talking about the mid-nineteenth century, "With the mineral revolution, Southern Tswana, already schooled by the civilizing mission in bourgeois ideas of property and progress, would learn the lessons of colonial capitalism first hand."[32] What was the form of this capitalism?

As colonial South Africa developed in the late nineteenth century, mining became a highly efficient engine of growth, as it overcame problems of labor shortages first by importing Chinese labor and then, when that became politically untenable, replacing Chinese workers with black laborers drawn from South Africa and the region.[33] The manufacturing sector was extremely limited, with the exception of "mechanical engineering concerns that met some of the needs of the mining industry and there was the dynamite factory."[34]

Why dynamite? No matter what is being extracted, explosives play an important role in the mining process. A leader of one major mining company described the history this way: "To be a big mining house you need lots of mines and to have lots of mines you had to be prepared to spend tens of thousands of pounds on holes in the ground looking for them."[35] Dynamite, as a revolutionarily powerful and transportable explosive in the second half of the nineteenth century, can thus be understood as primary material for the expansion of the mining industry.

For the mining companies, the creation of holes in the ground was also intertwined with the creation of complementary businesses. As the same mining executive explained, the reliance on making holes in the ground to find new mines is financially risky: "that's money written off the moment it's spent. So they decided very soon that they needed to diversify so that they could start businesses based upon local raw materials, satisfying local needs, and raising money for those needs on a booming stock market."[36] Here, "local needs" did not likely refer to the needs of the local *population*. It could refer essentially to vertical integration, because these industries often met the needs of the mining industry itself. Alternatively, the spin-offs might create and serve adjacent industries. Beginning spin-off industries using local raw materials such as coal tar and diversifying into industries such as dyestuffs could both mitigate the risk of the central primary production of mineral resources and provide a place to put available cash.[37]

Side industries could even become a way to dispose of waste products: at the AECI factory in Modderfontein, managers would find that the most efficient way to dispose of industrial effluent was to spread it on pastures as fertilizer.[38]

The patent and production evolutions of dynamite at the turn of the twentieth century can inform the understanding of problems around pharmaceuticals at the turn of the twenty-first: issues of profiteering, quality control, government concessions, and the challenges of manufacturing. As described in a historical account of AECI,[39] in the 1880s there was tremendous demand for explosives in South Africa to feed the growing mining industry, yet the South African government was concerned about "native" access to firearms.[40] The government sought to control production and gave a concession to one entrepreneur to establish a factory to produce dynamite locally, for perhaps multiple reasons: local production could mitigate the disruption of supply amid the wars over control of Africa; it would be part of making the colony self-sufficient; and it could lower costs.

However, the dynamite concession did not work out as planned. It went to a friend of Paul Kruger's, president of the South African Republic in Transvaal. The entrepreneur, perhaps emboldened by that relationship, ignored the requirement to produce locally and, instead, imported dynamite that was produced at Nobel factories in Europe and sold it at a huge markup.[41]

Not only were excessive funds demanded from the mining industry, but the entrepreneur also set things up in such a way that the profits went abroad. As historian Nancy L. Clark describes, he soon transferred controlling interest to foreign companies in partnership with the Nobel dynamite company and acted as their "front man" for operations in South Africa, which served both him and his foreign partners well: "selling imported dynamite at markups averaging between 100 and 200 percent, the concession earned profits in excess of £600,000 in 1897 and 1898, the bulk of which went to overseas shareholders."[42] Fundamentally, there was a mismatch between world-leading consumption of dynamite and affordable access to its supply. "Although the Witwatersrand [the region in which Johannesburg is situated] consumed half of the world's production, because of the monopoly granted by Kruger all of it had to be imported at prices far higher than local producers would have charged."[43]

At the time, the whole scheme was derided as a "barefaced swindle," but ending it took a while.[44] Eventually, the concession was canceled and a government-run factory began operations in Modderfontein. Most ingredients were still imported, but local production had its advantages. And even if profits were smaller than under monopoly rackets, they were more stable.[45]

The South African state's leading role in the country's turn-of-the-twentieth-century dynamite industry is an important example of what would become its model of "state corporations." The role of government in fostering mining and ancillary businesses was ever present, even as the state itself changed (as the Boer-controlled South African Republic in Transvaal came under control of the British Empire in 1902, and the Union of South Africa formed in 1910). African Explosives and Industries (which would later become AECI) was registered as a company in 1924, formed by a merger between the South African interests of Nobel Industries and the manufacturing arm of De Beers Consolidated.[46] The mining industry would grow tremendously during the 1930s and 1940s and would be the key source of wealth on which the apartheid system, put into place in 1948, was built. State-sponsored corporations would become key sites in which the state propped up the white minority—in the colonial period, under apartheid, and arguably well beyond.

It's worth returning to the detail that AECI was formed by the combination of the South African interests of Nobel Industries and the manufacturing arm of De Beers Consolidated. De Beers is among the most globally iconic South African brands, and the diamond company has its own deep ties with colonialism and apartheid. Its founder, Cecil Rhodes, was a key colonialist whose legacy has become a flashpoint for social movements to decolonialize education and culture in South Africa, most emblematically around the "Rhodes Must Fall" protests in 2015 that led to the removal of a statue of Rhodes that had been prominently positioned on the campus of the University of Cape Town.[47] Yet Rhodes's legacy is as mineral and industrial as it is social and political. As historian William Kelleher Storey has argued, "Rhodes was a visionary leader in business and politics who promoted advanced mine engineering while at the same time pressing for monopoly capitalism and racial discrimination, the sociotechnical imaginary that emerged in late nineteenth century South Africa."[48] Rhodes used the capital that he had accumulated in diamond mining in the Cape and Kimberley to develop a highly effective system for financing and managing gold mining in the Witwatersrand, setting the terms of the sociotechnical landscape of Johannesburg and of South Africa.[49]

South Africa's mining sector was strong globally not only because of its natural resources but also because of the innovations supported by combined and coordinated public- and private-sector investments (what we would now call public-private partnerships). In nationalist histories of mining in South Africa, as historians and science and technology studies scholars Paul Edwards and Gabrielle Hecht have pointed out, the emphasis is not on the presence of mineral resources but on the innovation by (white) South Africa. In their account of the rise of the Cold War–era uranium industry and how it invoked histories of the gold industry, Edwards and Hecht describe "narratives of metallurgic nationalism" that "did not dwell on the mere presence or extraction of uranium ore. After all, the former was an accident of nature, and the latter was performed by black labour. Instead, they celebrated (white) metallurgical skill, which had made South African mining great in the past and would do so in the future."[50]

Because of the long and deep history of mining innovation serving an oppressive regime and the white minority, it can be seen as un-African. Arguably, there is also value in appropriating that technoscientific legacy. What the philosopher Valentin-Yves Mudimbe has argued with regard to social science is just as true with regard to chemical science: "[I]t cannot be inferred that Africans must endeavor to create from their otherness a radically new social science. It would be insanity to reproach Western tradition for its Oriental heritage. For example, no one would question Heidegger's right to philosophize within the categories of ancient Greek language. It is his right to exploit any part of this heritage. What I mean is this: the Western tradition of science, as well as the trauma of slave trade and colonization, are part of Africa's present-day heritage."[51] In this chapter, I have only scratched the surface of the imbrications of mining and colonialism in South Africa. Given the specters of this history, how might synthetic chemists position themselves as heirs to this legacy of scientific innovation even in the face of its undeniable deep historical ties with settler-colonial violence, but this time using technoscience to serve multiracial democratic South Africans' own needs? What would it mean for the site of the manufacture of this "damnable invention" to become the site of innovative treatments for TB, HIV, and malaria? Might that offer a vision of turning swords into plowshares?

Parallels with the Hope for Antiretroviral Manufacture

The South African government played a key role in facilitating both the fully realized dynamite factory in Modderfontein, by providing the concession and then taking it over, and the fledgling pharmaceutical company at the same site, by providing start-up funding. State corporations such as Sasol have been emblematic examples of this model of scientific success through state-driven innovation in South Africa.[52] Applying the government-fostered or government-run industrial model to pharmaceuticals has many appealing aspects.

If the state were to decide to invest as much in fostering local pharmaceutical manufacturing as it had invested in early dynamite manufacturing, those efforts would face similar challenges. Take, for example, the case of local production of antiretrovirals. Just as it was for dynamite in the late nineteenth and early twentieth centuries, South Africa was one of the world's leading consumers of antiretrovirals in the late twentieth century and remains so in the early twenty-first century. Early in the antiretroviral era, in the late 1990s, US and other rich-country patents were enforced in South Africa, which made these drugs far too expensive to access. Strict intellectual property laws meant that South African drug prices were much higher than in other places, both developed countries and developing ones: in 1996 South African drug prices as a percentage of gross domestic product were among the highest in the world.[53] As will be discussed further in the next chapter, activists successfully fought for access to generic drugs produced in India and other countries with less stringent intellectual property laws. Allowing the importation of generic drugs from other developing countries—a practice known as "parallel importing"—was by far the largest step in making the drugs affordable in the country, lowering the price of antiretroviral therapy from thousands to hundreds of dollars per patient per year.[54]

Yet even at the time of this writing, antiretrovirals are not completely manufactured locally in South Africa; they are only formulated with imported ingredients. This has meant that South Africa has had little ability to control price and that South African companies and the state are reliant on decisions made in global markets for access to supply. South Africa has done better with these negotiations in the second decade of this century, with the public sector now paying rates that are low by international standards. However, its supply has been dangerously erratic, with "stock outs" in many regions interrupting care.[55] Moreover, as the South African currency has depreciated over the past few years, generic companies in South Africa (whether Indian controlled or just Indian supplied) have renegotiated pharmaceutical tenders that they had signed with the South African government, and the government had no choice but to pay up. If Indian pharmaceutical companies were to de-emphasize or even abandon the affordable antiretroviral market to prioritize products for other diseases and other parts of the world, South Africa would not have the capacity to meet those needs itself.

One anxiety around trying to manufacture pharmaceuticals locally would be the possibility of corruption, which as we saw, was an issue in the dynamite concession. How to make sure that those contracted by the state to supply pharmaceuticals really would make them locally rather than simply diverting the funds?

Another concern is how to make the revenue generated by pharmaceutical production benefit the local economy rather than just go to overseas shareholders. This is part of what drives the idea of state involvement, parallel to the state corporations of earlier eras, but for a democratic era. Whereas state corporations in earlier eras essentially benefited only the white minority, the hope here would be that a state role in pharmaceutical manufacture would benefit the society as a whole. However, this is complicated in a neoliberal era, in which the state's purview is more circumscribed, and yet it is the locus of considerable hope.

Any aspiration for building up industry and technoscience in South Africa is noninnocent: the material legacy on which iThemba sought to build in Modderfontein is inescapably entwined with colonialism and environmental destruction.[56] Any building on the history of dynamite would be a repetition with a difference, for good and for ill. Sociologist David Matsinhe has observed, "It is necessary to recognize that the reinvention of South Africa since the 1990s is not *creatio ex nihilo*. It certainly does not occur in a historical vacuum nor constitutes a complete break from the past. South Africa is reinvented out of the social, economic and psychological debris of its violent history that are (un)intentionally and (sub)consciously integrated in the new and emerging (reinvented) product."[57] Of course, iThemba's goal was even more ambitious than simply to make a widely needed commodity within South Africa in an economically sustainable way. They hoped to also use such resources to fund innovation, as state corporations in the apartheid era had successfully done. Making things and innovating things were intertwined. But it would be a repetition with a difference: rather than draw on the state to build a destructive and exploitative industry benefiting the few, they would draw on the state to build a reparative industry benefiting the many.

Contemporary Mining as Context of iThemba's Work

Today, mining continues to dominate the chemical industry in South Africa. In this context, chemists in all fields are aware of the mining industry as a possible career option. For those trained in the area of organic synthesis, this can make the mining industry seem like a site of a sector brain drain. For many, mining work is an undesirable option, failing to exploit their creativity and specialized skills. Synthetic chemists in South Africa can choose between a few fields: moving into chemical engineering in mining, doing formulation for the production-oriented firms, be they in paint or drugs, joining the tiny ranks of academia, or going into a general analytic field such as banking or consulting. Many consider emigrating to Europe or North America. At present, the mining industry dwarfs the pharmaceutical industry in South Africa. The tiny footprint of iThemba Pharmaceuticals on the sprawling land of AECI renders a concrete edificial illustration of the relative sizes of the industries.

The mining industry also takes the lion's share of resources, such as electricity, that might otherwise go to the broader population or to other endeavors. Access to electricity for majority-population residential consumers has long been highly politicized in South Africa—for example, in nonpayment protests during the apartheid era and service delivery protests more recently.[58] But industrial areas generally and iThemba's high-priority area especially are well served by the grid. Historian Nancy L. Clark has argued that from the beginning of the parastatal provisioning of electricity in South Africa, "[i]t was clear that electricity generated with coal from the mining industry would be returned primarily to that industry and any that served the mines rather than to the broad population."[59]

But mining broadly is also an important part of the political landscape, including in ways that cause the superexploitation of the colonial and apartheid past to haunt the post-apartheid present.[60] In August 2012 there was a massacre of platinum mine workers by police that for some revealed the "collusion of the state, mining capital, and sections of the labor movement."[61] The Marikana massacre was the most lethal use of force by the South African security forces against civilians since the end of apartheid—indeed, since the Soweto Uprising in 1976. In a dramatic confrontation, miners were gunned down by police: some in the back, some far from the front lines. Thirty-four were killed, and more than twice that number were injured. The event was shocking to many South Africans, but as historian Keith Breckenridge has pointed out, the massacre that initially "seemed in so many ways to be a surreal flashback to a repudiated past" came to be seen as "a symptom of the preservation of deeply formed structures of politics within the mining industry."[62] I would suggest that part of the allure of the dream of building a robust pharmaceutical industry in place of a mining industry is the desire to find a way for this violent history of struggle and repression to be overcome.

This tragic incident underscores a broader disillusionment with the capacity of mining to lead to broader social and economic development. This disillusionment is larger than South Africa: the challenges that mining-based economies present to developing democracies are being confronted all over the continent. For example, as anthropologist James Ferguson has explored, the economic boom of the 1960s and 1970s in Zambia was fueled by an extraction economy, mining copper and other minerals.[63] Ferguson tracks the disillusionment of urbanizing Zambians as the progress that they had been optimistic about stalled and reversed. Indeed, significant numbers of the Zambian middle class emigrated to South Africa to seek opportunities there, and Zambians are well represented in South Africa's scientific workforce, including at iThemba.[64] If mining will not be the location of a basis for development, there is a desire for alternatives. Why not pharmaceuticals?

In the Shadows, Flickers of Light

It is in the shadows of the profoundly fraught legacies of dynamite that scientists at iThemba sought to build a new scientific space for a new South Africa. The flickers of hope on this ground provide a distinctive route into the consideration of the possibilities of a postcolonial chemistry. This hope contrasts with the more monolithically destructive articulations that have been prominent in postcolonial and decolonial science and technology studies literature, in which the presence of chemical industries has been associated more unilaterally to exposure to harm—at such places as the site of the Union Carbide disaster in Bhopal, India,[65] and Chemical Valley near Sarnia, Canada.[66] There are important resonances with the hope that toxicological science holds in Dakar, Senegal, as described by Noémi Tousignant, in which measurement and control of toxicity serve postcolonial scientists' "struggle for capacity."[67] And yet iThemba's model is distinct because it seeks to start elsewhere. For these scientists and their networks of advisers and funders, the palpable and evocative history of chemistry at Modderfontein was not centrally about exposure to toxic chemicals but about a technoscientific legacy that served oppressive ends in the past but had the potential to be redirected to serve the people.

THREE

Science for a Post-apartheid South Africa

On December 5, 2013, Nelson Mandela died at the age of ninety-five after a long illness. He died at home in Houghton Estate, a leafy Johannesburg neighborhood just next to the one where I was staying. As the news spread, that small piece of the quiet neighborhood would become a festive site for mourners paying their respects.[1]

I was scheduled to fly to Cape Town the next morning to visit H3D, a drug discovery effort operating out of the University of Cape Town, to meet with a former iThemba Pharmaceuticals scientist and their new colleagues. I allowed extra time to get to the airport. All was peaceful, and the flight uneventful. It was a good time for reflection, on questions large and small. That morning, I thought back on how inspiring Mandela had been from afar: for example, when watching his 1990 release from prison on television as a teenager in the United States. I also pondered more proximate connections: how had South Africa's transition from apartheid to democracy shaped the trajectories of pharmaceutical science there?

The lunch that day with the scientists at the University of Cape Town Club was delightful, brimming with dynamic conversation about the possibilities for drug discovery in South Africa. It was cut only a little bit short, because H3D's director said that he would like to attend the campus memorial for Mandela. We walked up the hill to the university's Upper Campus together and stood in the brilliant sunshine listening to administrators eulogize the great man who had been such an important leader in South Africa's anti-apartheid struggle and in the transition to democracy. We joined a multiracial crowd of students, faculty, and staff gazing up at esteemed speakers, amid beautiful academic buildings perched alongside Table Mountain. Science per se didn't come up that day, but during his presidency, Mandela had argued forcefully that science and technology are a bellwether for a nation's social and economic development, and that the strength in science and technology that had been developed in South Africa during apartheid could and should be transformed to serve the newly democratic society.[2] In profound ways, the lives and the life's work of the Zambian lab director and the South African research scientist whom I was visiting exemplified the effort toward realizing that dream.

Early the next week, back in Johannesburg, I attended Mandela's funeral service in a soccer stadium in Soweto. There was, uncharacteristically for the season, a cold, driving rain. My train was waylaid in the rain: electricity problems. Yet the atmosphere on the journey and then at the stadium was jubilant, the crowd singing struggle songs and thrilling to the international guests, who included US president Barack Obama and United Nations secretary-general Ban Ki-moon.[3] The presence of world leaders was deeply symbolic: during apartheid, South Africa had been cut off from the world, but it had become embraced in its transition to democracy. Nelson Mandela was an incomparable figure in South Africa, as a freedom fighter, as a head of state, and, even well into his postpresidential years, as a moral leader.[4] Yet his ambitious vision of a just, multiracial democracy has not been fully realized.

Mandela's vision and legacy and the broader successes and failures of South Africa's transition to democracy all contributed to shaping iThemba Pharmaceuticals and its mission of finding new drugs for TB, HIV, and malaria. The nineteen years between Mandela's ascension to the presidency and his death had been formative ones for the small pharmaceutical company at the core of my research. Over the course of my fieldwork at iThemba, the death of Nelson Mandela was among many events that occasioned broad public discussions of the progress and shortcomings of the still-developing South African democracy, including among the scientists involved with the company. In the push and pull of hope and disillusionment in the young country, I would come to realize that the period in which I started my research, during the lead-up to South Africa's hosting of the men's 2010 FIFA World Cup, marked a relative high point for hopefulness in the country.[5] The company's name, which is Zulu for "hope," was thus fitting for its time.

The moment of South Africa's democratic transition is generally counted from the 1994 election of Nelson Mandela as president, but of course the transition began earlier,[6] and it meaningfully continues today. The first democratic election was within the living memory of all involved with iThemba, and many of the country's founding leaders were still public figures and touchstones. This transition from apartheid to democratic rule is an important part of how iThemba can be historically located. Colonialism and apartheid created the context of the disease burdens that iThemba sought to solve, but it also created the conditions of possibility of doing drug discovery and development in South Africa. The emergence from apartheid at the same time as the ascendance of HIV and structural adjustment created a situation for South Africa in which there was a great deal of hope that a more just and equal society might be created—including with regard to health—but routes toward progress were very constrained.

In this chapter, I will draw together three historical threads from the 1990s and early 2000s that may seem disparate: the landmark court case over South Africa's right to import generic drugs, dubbed "Big Pharma versus Nelson Mandela"; South African scientific exchange initiatives that aimed to foster a multiracial scientific workforce for the new South Africa's R&D capacity; and efforts by scientists outside South Africa to bring the young democracy into the international community of drug discovery. These threads highlight interrelated sets of tensions: science in a democratic South Africa as both discontinuous and continuous with apartheid-era inequalities; South Africa as both developed and developing; and R&D capacity as simultaneously local and transnational. A broader tension that underlies these is South Africa as both a nonracial and a multiracial society, democratic and unequal, and the country's position in Africa and in the world at large. I will conclude the chapter with a discussion of Nelson Mandela's own aspirations for the role of science and technology in the new South Africa. iThemba manifested the challenge of working within these tensions and can be read as an endeavor toward realizing Mandela's vision of science in the service of the people.

Access to Antiretrovirals for Democratic South Africa: Big Pharma versus Nelson Mandela

The copresence of strong intellectual property (IP) laws and demands that pharmaceuticals serve the people is an important part of the post-apartheid landscape that shaped iThemba. Among those involved with iThemba, South Africa's strong IP laws were one of the most frequently noted advantages to doing drug discovery in South Africa as opposed to other developing countries. This legal framework might be counterintuitive to international observers, since South African activists and their supporters in the young government famously challenged global pharma for access to generic antiretrovirals in the late 1990s—and won.

Contestation over pharmaceuticals has been part of the struggle over the scope of South Africans' inclusion in the benefits of global science since early in the democratic era. How would the companies based in the country be positioned as South Africa became more deeply integrated with the globalized knowledge economy, and how would the new democracy's citizens be positioned in global hierarchies of access to pharmaceuticals? To what degree would these desirable products of global technoscience be made and be accessed—by and for the privileged minority, and by and for South Africans as a whole?

The years around the 1994 elections also saw the emergence of a crisis of HIV/AIDS in South Africa, which would come to have the highest number of HIV/AIDS cases in the world. Effective treatment became available in the United States and other places in 1996. Social movements, especially the Treatment Action Campaign, built on the success of the anti-apartheid struggle to again turn to the international community, this time to mobilize against South Africans' exclusion from the treatments that were transforming the lives of those with HIV in rich countries.[7]

This struggle for access to drugs for HIV led to high-profile conflict between South Africa and the United States at state and corporate levels. According to sociologist of neoliberal biomedical economies Melinda Cooper, "Beginning in the mid-1990s—the watershed decade that saw the end of apartheid and the introduction of antiretrovirals (ARVs)—the United States has consistently tried to dissuade South Africa from using the emergency clause that would allow it to override the World Trade Organization (WTO) rules on importing low-cost generic drugs."[8] Nevertheless, in 1997 the South African government passed the Medicines Law, which allowed for the importation of antiretrovirals from generic producers in countries with less protection for IP, notably from India.

In February 1998 a group of pharmaceutical manufacturers—the South African Pharmaceutical Manufacturers Association and thirty-nine mostly multinational companies, supported by the US government as represented by Vice President Al Gore—sued the Nelson Mandela government, arguing that the Medicines Law violated the South African constitution and the international Trade-Related Intellectual Property Rights (TRIPS) Agreement. Paving the way for a public relations disaster for Big Pharma, the case was widely dubbed "Big Pharma versus Nelson Mandela."[9]

In this court case, it's notable that South African pharmaceutical manufacturers were on the side of Big Pharma and against South Africa's new majority-rule government. This underscores internal divisions within South Africa that carried over from the apartheid era. There had been a local pharmaceutical industry in the apartheid period, which had been relatively small, generally only formulated drugs from imported ingredients, and focused on serving the South African elite. That legacy industry had long cast its lot with transnational pharmaceutical companies[10] and, in the period of transition, did not shift its allegiance to the interests of the newly empowered South African majority. The South African pharmaceutical industry's longstanding service of the interests of global capital and provision only to the elite meant that it was not in a position to manufacture antiretroviral drugs locally anyway. At the time of the transition to democracy, 20 percent of the population, mostly white, was covered by the private health care system, and that minority accounted for 80 percent of the total drug expenditure.[11] Many powerful interests, globally and domestically, had a stake in maintaining South Africa's two-tiered system.

South Africa was in a globally unusual position with regard to international property protections under TRIPS, since it was a developing country by most economic measures but had a developed-country IP regime. South Africa's IP regime can be seen as a legacy of minority rule: South Africa was a founding member of the landmark international trade agreement the General Agreement on Tariffs and Trade in 1948,[12] the same year as the official implementation of apartheid. For the white minority, South Africa was (and is) a developed country. South Africa (unlike other developing countries) lacked the flexibility to manufacture generic versions of patented drugs without the permission of the US and European patent holders. It would have had to follow a more legally onerous path of "compulsory licensing." India's greater TRIPS flexibility was an important reason that Indian companies were able to make generic antiretrovirals so soon after branded versions were released. Thus, South African activism was advocating for parallel importing from India rather than for making the drugs themselves, and parallel importing was the path that the government pursued.

The dominant discourse of the impact of TRIPS on resource-poor countries paves over the considerable heterogeneity within the Global South.[13] Developing countries had until 2006 to comply with protecting the IP of developed countries,[14] and least-developed countries had until 2016, but South Africa was obliged to comply immediately. Indeed, many believed that South Africa "was being held to a 'TRIPS plus' standard (a higher level of patent protection than required by TRIPS) both by the U.S. government and by the private plaintiffs in the lawsuit."[15] The 1997 Medicines Law sought parallel importing under article 31 of TRIPS, which allowed any member country to do so in cases of emergency. The scale of the HIV/AIDS crisis in South Africa—with at least 16 percent of the population infected at that point, including 20 percent of pregnant women—certainly seemed to qualify as a national emergency,[16] and South Africa was seeking, not to undo TRIPS or to be excluded from its obligations, but to utilize an article of the agreement.

Although none of the iThemba cofounders were directly involved in the Big Pharma versus Nelson Mandela case, some paths intersected—notably that of David Walwyn, who served as the chief commercialization officer for iThemba at the company's height and who met Big Pharma executives in 1999. At the time, he was responsible for commercialization and IP at the South African Council for Scientific and Industrial Research, and he attended an investment conference in New York to make a presentation on opportunities in the South African pharmaceutical sector. At lunch, Walwyn was interrogated about the court case by the head of the trade group Pharmaceutical Research and Manufacturers of America and the head of Merck, and when he indicated that he thought that the pharmaceutical companies might lose, "I think we all choked on our lunches, me from terror and the bigwigs from disgust."[17]

In 2000, in the face of the public relations disaster, the US government ended its threat of sanctions over the question of parallel importing. President Bill Clinton issued an executive order that "the United States shall not seek, through negotiation or otherwise, the revocation or revision of any IP law or policy of a beneficiary sub-Saharan African country . . . that regulates HIV/AIDS pharmaceuticals or medical technologies," and prohibited the US government from taking actions that would block access to AIDS drugs.[18] In 2001 the industry plaintiffs finally acknowledged defeat and dropped the "Big Pharma versus Nelson Mandela" case.

The year 2000 also saw the World AIDS Conference hosted in South Africa. While the turn of the twentieth century and the recent legal victory might have been celebratory moments for the country, they occurred in the context of serious challenges and attendant discord and division among South African state actors.[19] At that point, as historian Mandisa Mbali has pointed out, there were two challenges facing treatment access for South Africans, one global and one domestic: "the prevailing consensus among donor countries was that it was not 'cost-effective' to provide ARVs [antiretrovirals] in resource-poor settings," and "the country's president, Thabo Mbeki, had recently denied the basic science behind the disease and the efficacy of ARV therapy."[20]

As other analysts have argued, these two challenges should be seen as interconnected: the prohibitive cost of antiretrovirals could make provision of them by the young country seem too daunting, regardless of their efficacy.[21] The legacy of a history of apartheid and the dependent position of the young democratic but neoliberal South Africa on global pharmaceutical networks for access to antiretrovirals contributed to HIV denialism there: under conditions in which the cost of the drugs meant that they could not realistically be provided to all South Africans, skepticism about their efficacy provided a way for government actors and others to articulate the lack of access to drugs as something other than failure.[22] The struggle for access to antiretrovirals and skepticism about their efficacy both became part of the post-apartheid legacy that iThemba inherited.

Like the anti-apartheid movement before it, the movement for access to antiretrovirals involved a combination of domestic and transnational activism.[23] Yet activists were not alone in combining domestic and transnational efforts to try to shape post-apartheid South Africa's future: scientists, including those who would become involved in iThemba, were working across these scales as well, with their own set of hopes.

The turn-of-the-century struggle for access to antiretrovirals set up the discursive terrain of pharmaceuticals in post-apartheid South Africa as one principally about IP and access to imported drugs. However, there were also parallel developments in knowledge infrastructures that would lead to the effort to bring South African drug discovery into the scope of pharmaceutical imagination.

Building Multiracial South African Science through International Exchange Programs

Although iThemba's founding took place in the twenty-first century, formative parts of its prehistory can be located in the 1990s. The iThemba cofounders' visions for building science in South Africa were informed by their visions for the future of South Africa itself. This is well aligned with the concept of sociotechnical imaginaries: ideas about what science should be are coproduced with ideas of what society should be.[24] This necessarily includes ideas about race and exemplifies a persistent vacillation in the understandings of both science and the South African nation as alternately *nonracial* and *multiracial*. Nonracialism and multiracialism have long been copresent in South African antiracist discourses, including within Nelson Mandela's party, the African National Congress.[25] Nonracial discourses can sometimes act as a blinding white light and can sometimes be mobilized as part of a neoliberal agenda, but their imaginative work can also be more expansive.[26] Multiracial discourses have their own promises and perils. How have nonracial and multiracial imaginaries been at stake in scientific research and development, including at iThemba?

One of iThemba's cofounders, Frank Fisher, is a white South African who was a senior figure in South African R&D science during the democratic transition. Fisher spent the decades of the 1970s, 1980s, and 1990s at African Explosives and Chemical Industries (AECI) and the major mining company the Anglo-American Corporation Group. In an interview, Fisher enthusiastically characterized himself as "an African" and described that identity as his reason for returning to his country after working for a few years in Scotland:

> I graduated from the University of Natal, and I went to work in Scotland for a military research establishment in Scotland, working on explosives. And I met my wife in Scotland. But I realized I was an African. You know I could speak English, Afrikaans, and Zulu, and I would often think to myself in the pouring rain in Scotland, what the hell am I doing here? I really should be in Africa. So we came back to Africa, and I joined a company then, the largest chemical company in this country in those days. It was called AECI. It stood for African Explosives and Chemical Industries. Part of the ICI Group, the British ICI Group. Their job was to make explosives for the mines here in Johannesburg. The largest explosive factory in the world was here in Modderfontein.

Fisher had built his career during apartheid, but he also participated in renegotiating notions of what belonging in South Africa could mean for white South African scientists like himself, as well as for others. His work to build a multiracial scientific workforce would mark a shift akin to what Sarah Nuttall has characterized as "belonging separately, as under apartheid, or belonging together (sharing) as in the postapartheid context," in which white belonging as it had been in the settler state could no longer be assumed but needed to be negotiated on new terms.[27]

At the height of the democratic transition, Fisher was in a leadership position at AECI, and he told me in an interview in 2010 that "in '96, '97, the board of the AECI put pressure on every manager to start employing blacks." If R&D were to continue to be supported under the new regime, those involved would have to be more inclusive. Nelson Mandela's government was broadly pro-science, and diversification of the scientific workforce would be important for securing broader democratic buy-in. This is an affirmative action logic, but one that plays out differently from that in many other contexts, because the imperative was to include the newly empowered majority. In an affirmative action logic in the United States, for example, we might say that what is being *affirmed* through *affirmative action* is the full citizenship of African Americans and other minority groups. But in the affirmative action logic that Fisher was describing, what was being affirmed was science itself. The demographics of the scientific workforce in apartheid-era South Africa had been an element of why science itself was seen as tied to minority rule, and a more inclusive workforce would affirm that science could have a legitimate role in the new majority-rule polity.

Fisher agreed that hiring black PhDs at AECI was a fine idea, but there were obstacles. There were very few black PhD scientists in South Africa at the time, and Fisher said that he was unsatisfied with the quality of the black PhD scientists produced in South Africa, which he saw as a legacy of the apartheid education system. Fisher explained that during apartheid, blacks had to go to separate universities for their undergraduate degrees, and black-serving educational institutions didn't train PhDs. There were a few black PhDs trained at then-white institutions such as the University of Cape Town, but "the staff of the University of Cape Town, generally speaking they didn't expect blacks to do as well as the white guys. It was all sort of accepted that he wouldn't be quite as good. And that bothered me." He himself had been trained in the United Kingdom, and he believed that experience at an international university where high expectations were set for all students—black students included—would help to improve the quality of the pool of black scientists that he could hire. In this account, cosmopolitan UK universities are figured as a way to train South African scientists outside parochial South African racial hierarchies.[28] Exposure to what are imagined to be *nonracial* global standards of science in the United Kingdom would help these young scientists to participate in building the same thing in South Africa.

For Fisher there were global means to his very local ends. Motivated by the need to hire black PhDs so that scientific research would be supported in South Africa's democratic era, Fisher sought ties with international scientists who would assist in their training, and he worked with them to set up exchange programs for PhD students. These exchange programs focused on strengthening the education of black PhDs, and they also deepened the relationships among the UK- and US-based scientists who would go on to become the other cofounders of iThemba. This in turn would be a way to realize a vision of creating inspiring work opportunities for diverse South Africans in R&D.

The first exchange program that Fisher organized was with a longtime

AECI consultant, Professor Tony Barrett from Imperial College London. As Barrett told me in an interview at his university office in London in 2014, he first went to South Africa in 1982—as he put it, "before it was fashionable." This was a coy way of describing his having consulted for the apartheid government, and yet it also positioned him well to collaborate with post-apartheid initiatives. Barrett had spent ten years on the faculties of US universities in the late 1980s and early 1990s, visiting South Africa from there. He returned to Imperial College London in 1993 and became a consultant for AECI, visiting South Africa frequently. The timing of his return to Imperial College was fortuitous, because Barrett had a ready collaborator in Julian Walsh, a former trader in the oil industry who had retired early to work on building public-private partnerships at Imperial College in 1992. Walsh would prove to be a highly enthusiastic coconspirator in exchange programs and in the founding of iThemba, exhibiting a combination of philanthropic spirit and commercial mindedness.[29]

Fisher set up an exchange program with Barrett's lab in London in the 1990s, in which black students in PhD programs at South African universities would spend a year at Barrett's lab. The program was a success, which was not very surprising to Fisher, because if "you put the best students in a really good university, they are going to do quite well." Fisher retired from AECI soon thereafter, but he believes that the students trained then are still doing well, many now in senior positions in South African government and industry.

It was Barrett who suggested expanding the exchange program to his friends in Atlanta, specifically the brothers Charlie Liotta at Georgia Tech and Dennis Liotta at Emory. Barrett's argument for this expansion was notable with respect to post-apartheid nation building: Barrett suggested that the presence of large numbers of affluent African Americans in Atlanta would have a positive influence on these black graduate students, and Fisher was receptive. In this way, the aspiration for a world-class black scientific workforce was linked to the aspiration for a black middle class and professional elite. This is a more explicitly *multiracial* frame—if British universities and cities were figured as preferred training locations because they were imagined to be less racist than South African ones, American and especially Atlantan universities could be even better because black excellence and leadership were so prominent in the city's social and political spheres. The project of training scientists to participate in R&D after apartheid was thus both a scientific and a social enterprise.

We could certainly problematize this figuration of Atlanta as a model of a place of black success. Integration and justice are certainly incomplete in

Atlanta. Yet at least since W. E. B. Du Bois's iconic imagining of the city in "Of the Wings of Atalanta" in his 1903 *Souls of Black Folk*, Atlanta has been figured as a beacon of multiracial hope.[30] Atlanta's position as the home of Martin Luther King Jr. makes it a particularly appealing analogue for hope for the city and country of Nelson Mandela. Although the dreams of Martin Luther King and of Nelson Mandela have been unfulfilled, they have important resonances with each other and remain powerful touchstones in their respective contexts and globally.

Even as Atlanta's areas of entrenched black poverty are occasionally featured in globally popular hip-hop music, Atlanta's contemporary black *prosperity* is more prominently broadcast, including in South Africa, through music and especially television programs such as *The Real Housewives of Atlanta*. Over the course of my fieldwork, when I introduced myself as based at Georgia Tech in Atlanta, many South Africans—especially but not limited to middle-class South Africans of color—marveled. In notable contrast with other black-majority cities in the United States, Atlanta is widely associated, not with black power, but rather with a nonmilitant black empowerment and racial coexistence—and of course, whites still hold major political and especially economic power, and so its model is not alienating to white South Africans. Many middle-class South Africans of all races have strong and hopeful associations with the city.

By the time that he was involved in the cofounding of iThemba, in the early 2000s, Frank Fisher also had a second goal, in addition to training black graduates: bringing back (white) South African scientists who had emigrated to Europe. Fisher had worked with many excellent scientists over the years at AECI, and he wanted them to come home to contribute to their country. Even as this hoped-for return does not seem to have come to pass, it may be that opportunities to do R&D work in South Africa have stemmed the brain drain somewhat, preventing some white South African scientists from leaving the country.

In his account and in his life's work, Fisher operates with plural racial imaginaries. A multiracial imaginary articulates a future of science that for the rainbow nation of South Africa would include both white scientists and black scientists. This coexists with an implicitly nonracial imaginary of simply training and retaining as many scientists as possible. US-style liberal "color-blind" ideologies might not be tenable in South Africa, but racial categories might be highlighted or de-emphasized in the service of universalist goals such as science.

To retain scientists in an economy, employment opportunities are vital. To this end, Frank Fisher played a role in successfully securing start-up funding from the South African government for the founding of iThemba. As discussed in the previous chapter about the legacies of dynamite, colonial and apartheid-era South Africa has an extensive history of state-owned enterprises that employed and benefited the white minority. This twentieth-century legacy created a situation in which it was reasonable to seek funding from the South African government for an enterprise that would benefit South Africans in the twenty-first century. However, because the post-apartheid state was avowedly market oriented,[31] the stage was set for hybrid models of public-private partnership.

Insofar as its roots are in trying to bring black South Africans into R&D, iThemba's founding can be connected to postcolonial efforts to *Africanize* science, in the nonacademic sense of the word as it is used in other regions in Africa, as in "Africanization of the civil services."[32] The fact that the company was 50.1 percent owned by the South African government makes a civil services frame particularly apt. And yet, as is common in post-apartheid South Africa, iThemba operated with plural senses of who counts as African: one sense specifically focused on the black-majority population of the country, and another sense was far more expansive and included black, white, and Asian South Africans, as well as (black) Africans from other countries in the region. At iThemba, as in South African science as a whole, black South Africans remained underrepresented relative to white South Africans and Africans from neighboring countries.[33]

There might be more capacious ways to think about Africanization, beyond those that align with the ideas of iThemba's cofounders. Bringing more Africans into science is a very circumscribed notion of Africanization, because the central idea is to bring Africans into an already-established field that is not anticipated to be transformed. There is little chance of Africanizing synthetic chemistry in a way analogous to what Lyn Shumaker described as "Africanizing anthropology": attentive to the process of fieldwork, Shumaker argues that African assistants and informants had a role in shaping anthropology's knowledge construction about them, and she highlights anthropological knowledge making as a collective process rather than an individual one.[34] Drug discovery is definitively a collective process, and iThemba's goal was to shape knowledge making about African diseases, but the vision was to do this work in accord with existing scientific paradigms. Drug discovery is a mature field of practice that would be hard for Africans to fundamentally re/shape—and none of those involved in iThemba were interested in doing so.

Yet a third sense of Africanization can help us to understand iThemba's project as well: Esperanza Brizuela-Garcia's discussion of the Africanization of African history as an endeavor for greater *authenticity* and *relevance*.[35] With its racially diverse staff, global networks, and marginal ties to indigenous knowledges, iThemba may present a weak case for *authentically* African knowledge—at least in the persistent conventional framings of "authentic African" and "modern" as mutually exclusive. Yet because its founding goals were to find inexpensive new drugs for TB, HIV, and malaria, it was certainly striving for *relevant* knowledge. This makes its project an example of democratic, post-apartheid science—iThemba's mission broke from science of the apartheid era by seeking to serve the South African majority.[36]

Bringing South Africa into the International Community of Drug Discovery

The decision to extend the training network of South African scientists to Atlanta was formative in setting the stage for a path toward founding a company with a drug discovery mission. South African cofounder Frank Fisher had been interested in maintaining and building scientific R&D infrastructure broadly, and he was not primarily concerned with which field would be pursued. His Imperial College collaborator, Tony Barrett, also worked across a range of scientific spheres. But Dennis Liotta, the scientist who would become Fisher's main Atlanta collaborator and a cofounder of iThemba, was and is centrally focused on drug discovery. A professor of chemistry at Emory, Liotta had come to global prominence because of the very same category of drugs that South Africans had been struggling to gain access to in the late 1990s: he and collaborators had discovered key drugs in the antiretroviral cocktails that had transformed HIV from an acutely mortal disease to a chronic one—at least for those with access to the drugs.

Liotta became involved in South Africa in 1999, through Tony Barrett, as part of a consortium for exchange programs. As told to me by Liotta in an oral history interview in 2012, he and his colleagues toured South Africa's research facilities, were impressed with the quality of science education, but were struck that there was very little drug discovery going on. He came to believe that this was a consequence of the former South African regime, in a couple of ways: first, sanctions that had been imposed on South Africa had precluded international pharmaceutical companies from putting R&D centers there, and second, South African scientists with synthetic chemistry skills didn't like the system, were relatively mobile, and left. Thus, his assessment was that there was "a whole generation . . . trained without knowing anything about drug discovery and development."[37]

Other analysts render the periodization of South Africa's isolation from the global pharmaceutical industry somewhat differently, suggesting that the major decrease in pharmaceutical R&D in South Africa happened in the later 1990s, as South Africa became less isolated and economically liberalized at the same time that the global industry became more consolidated, such that South African companies were increasingly affected by global competition and consolidation trends.[38] More broadly, apartheid-era South Africa's exclusion from global science was never so complete as the discourse of sanctions implied.[39] Be that as it may, by the turn of the twenty-first century, when Liotta was first exposed to the terrain of scientific research in South Africa, pharmaceutical R&D activity there was quite small.

Since drug discovery is such a specialized area, Liotta argued that it would take "a well-defined knowledge transfer program" to "fill that gap." He told me:

> So we immediately realized, wow, this is an interesting opportunity, because we, and the network of people that we knew, had the skills. And if we could find an effective way of transferring those skills, we could train the next generation of African scientists to address their own problems. And that just became an instantly compelling driving force for me. It was just like, wow. And I would talk to people, to scientists, and I would describe what we were trying to do, and literally to a person, they all said, "Let me know how I can help."

The late 1990s was a period of great global enthusiasm for South Africa's transition to democracy, including in the United States. There had been widespread fears of civil war, and many around the world were inspired by Nelson Mandela's leadership through a largely peaceful transition. The notion of contributing to this emerging society was appealing. However, according to Liotta, it was not without its complications:

> But in the process of doing this, we also came to this very profound conclusion, which is that if we did this, if we got people excited about this area, and gave them the skill sets they needed, but there were no jobs for them, then like many others who preceded us in Africa we would make this well-intentioned blunder of creating a brain drain in essence. So we decided that we'd have to, simultaneous with our knowledge transfer program, we'd have to do something to catalyze the formation of new commercial entities that could hire the people who were being trained and could then make those discoveries. And that was essentially the origins of iThemba Pharmaceuticals.

In Liotta's model, strong local university research matters: scientists based in South Africa are not empty vessels waiting to be filled with imported knowledge. The strong university training that Liotta saw is itself a legacy of colonialism and apartheid, built with the training of white settler colonists in mind. The presence of these historically white educational institutions made South Africa a particularly ripe place for this kind of knowledge transfer project, relative to other African countries that largely lack this robust research base on which to build.

At the same time, strong universities are not enough to create an environment conducive to drug discovery science. In Liotta's framing, both human capital and IP need to find space in the commercial sphere in order to be sustainable. This vision of potential synergy between government and for-profit entities to build capacity for pharmaceutical knowledge production is consonant with broader global trends in public-private partnerships for drug development,[40] but the foregrounding of employment in addition to markets is distinct. The provision of employment opportunities for southern African chemists is framed as a necessary precondition for the possibility of southern African pharmaceutical innovation. The approach of building up a sector with an emphasis on employment rather than on products produced or raw impact on economic growth is common enough across sectors[41] and is an appropriate topic for analysis of pharmaceuticals as well.

There is an open-ended sense in which building South African capacity in this area might be possible, but success is by no means predetermined. Science and technology studies scholars have tracked ways in which "brain circulation" can foster Asian bioscience rather than accelerate "brain drain,"[42] and it is an open question whether this model can work in the much less developed context of South Africa. This element of the unknown is precisely part of the appeal. The unknown points to the potential to "make a difference," and this is one of the reasons that when Liotta told his scientific colleagues about his hopes for fostering African scientific capacity, they were all so eager to help. Postcolonial science and technology studies scholar Itty Abraham points out that "postcolonial locations thus include relations of weakness and possibility, valences that cannot be known in advance but that are products of historically situated intersections of the political economy of place and unequal location within transnational circuits of knowledge flow."[43] The very contingent network that came together around iThemba inspired feelings of real opportunity, possibility, and hope.

The sense of possibility and hope that attracted these scientists in the

1990s resonated with the broader global community's affective investment in South Africa's success as a multiracial democracy. South African literary scholar Leon de Kock provides an example from a literary conference. He describes a German theorist speaking in Johannesburg in 1998, declaring, "It is imperative for us that you succeed!" For de Kock, "And therein lay the key. For many reasons, it is imperative for *others* that South Africans succeed at the democratic, multiracial miracle that *they* (the non–South Africans) have yet to see realized in their own countries. South Africa must carry this burden of moral example, just as in earlier times it carried the burden of having to be a moral pariah of the larger world."[44] There can be a sense that South Africa's success in building a multiracial democracy would show that that elusive goal is possible for other multiracial countries. Moreover, its success can seem imperative for showing the possibility of African success. One iThemba bench scientist from a country bordering South Africa described South Africa as "the flagship of Africa." I asked what was meant by this and was told:

> It's our model, it's our main project, I think. A lot of energy, not only from an African perspective. Even from, I think, the donor organizations, when they want to set up a company in Africa, I think seven out of ten times it will be in South Africa, or even eight out of ten. So they look at it as a country that can be an example. It's a young democracy. They had all these things in '94 going through reconciliation after apartheid. So they have a very good touch to what the donor world would consider a new democracy. And in that manner, I think that the donor of the funds would want to see this country as a project which should work.

Because of South Africa's role as the most developed country on the continent, the reverse is also true: the stakes of failure are very high.

> Well, I mean it matters for everybody in the sense that from an African perspective, South Africa could well be a model country, as the continent is coming toward development. If this country goes wrong, I think we will wonder, what now for the continent? Well, from the perspective of iThemba, if we can't do business here, imagine doing it in [my country]. It would be a hundred times more difficult.

I asked this scientist whether South African discoveries would be meaningful to other Africans or not, given South Africa's differences, and the reply was intertwined with symbolism:

> I think it would be meaningful. It would set an example to the rest of the African countries to say, South Africa can do it, anybody else can do it. They are hosting the World Cup now. Who thought Africa would ever host the World Cup? Other countries will in the future think, South Africans did it, we can do it. So I think same goes for the pharma industry. If South Africans can discover and develop drugs, then everybody else can do it. It's just a matter of having the will to do it.

In wide-ranging ways and from many vantage points, it can seem as if South Africa's failure would mean that highly desirable goals—multiracial democracy, African contributions to drug discovery—simply can't be achieved.

Dreams of Realizing Mandela's Vision of Science in the Service of the People

Dennis Liotta's account of the challenge and possibility of bringing South Africa into the world of drug discovery science resonates with Nelson Mandela's 1995 articulation of the role of science and technology in the democratic transition. Mandela lauded science and technology in general, arguing, "A nation's commitment to science and technology, or S&T, is often an indicator of its stage of economic, social, and cultural development." And yet even as Mandela boasted of South Africa's strength in the area, he was explicit about science and technology's fraught history in the service of a racist regime: "In South Africa, we are fortunate to have a strong S&T infrastructure, including numerous institutions and well-trained people. Unfortunately, these structures and people have been guided by a policy based on benefiting a minority and supporting the security needs of a regime whose primary objective was to maintain apartheid."[45] Mandela did not emphasize the racial noninclusiveness of South Africa's scientific workforce, but rather its goals in the service of apartheid. Building on South Africa's strength in science and technology while changing these goals would exemplify an agenda of dis/continuity.

If the project of science in South Africa could not be understood in isolation from its domestic political context, neither could it be understood in isolation from the international political order. In Mandela's articulation, as in Liotta's, global abhorrence of apartheid led directly to South Africa's exclusion from sharing in global innovation: "The use of S&T to bolster apartheid led to an international cultural and scientific boycott of South Africa. This had the effect of excluding South Africa from the international flow of S&T knowledge. A direct consequence has been declining investment and innovation in the area, which has had a negative impact not only on S&T research but also on the overall development of our nation." Thus, for Mandela as for iThemba's South African and international founders, local science was inextricable from transnational connections. Moreover, Mandela posited hope for South Africa as the locus of hope for Africa as a whole:

> I remember the advice given to me by one of the international members of the . . . African Academy of Sciences, about how important our future investments in the area of S&T are and how many African countries are only realizing now that they would have been further along the path of economic development had they recognized, following independence, the importance of investment in this area. This lesson has not escaped our new government and indeed S&T is recognized as a major pillar of the Reconstruction and Development Programme.[46]

For Mandela, the post-apartheid state had a role in fostering science as part of its broader mission. The South African government's financial support for iThemba exemplifies this promotion of science and technology as a route to economic development—and to democratic development as well. When the famously anti-science Thabo Mbeki succeeded Nelson Mandela as the second president of South Africa from 1999–2008, this support for science faltered in high-profile ways, yet it lived on in, among other places, the initiatives that would become part of iThemba's prehistory. When Jacob Zuma assumed the presidency in 2009, there was a great deal of hope—not least for reinvestment and recommitment to scientific research.[47]

The Nelson Mandela period was characterized by one kind of antagonism to Big Pharma: the demand for parallel importing, which came to a head in the court case "Big Pharma versus Nelson Mandela." The Thabo Mbeki period was characterized by another kind of antagonism to Big Pharma: HIV denialism and attendant skepticism of pharmaceuticals. Although there has been considerable disillusionment since, the Jacob Zuma period began with high hopes that there might be a realignment: a recommitment to science and with it a possibility that local innovation could work in concert with Big Pharma—this time serving the people of South Africa, in harmony with the continent and the world.

FOUR

"African Solutions for African Problems"

"African solutions for African problems." This was a slogan that came up again and again in my conversations and interviews with drug discovery scientists at iThemba Pharmaceuticals, the small start-up company in the outskirts of Johannesburg founded in 2009 with a mission of finding new drugs for TB, HIV, and malaria. South Africa is, of course, a very problematic stand-in for the African continent as a whole, but "African" does important work. That moniker is also an actors' category, routinely invoked by the scientists themselves, and I am following their lead in adopting it provisionally as a category of analysis even as I do not take its constitution or its boundaries for granted.[1] As we will see, the term "African" is remarkably flexible, able to incorporate South Africans of diverse ethnicities, as well as (black) Africans from other parts of the continent who are working in South Africa.

As we have seen in earlier chapters, the scientific advisers to the company were trained at places like Cambridge in the United Kingdom, and the majority of those elite scientists continued to be based in rich countries. However, the bench scientists at iThemba all had meaningful African roots and strong South African ties. The number of bench scientists at iThemba ebbed and flowed, peaking at nearly twenty but often closer to ten, and their demographic composition varied. They included white, Indian, and black South Africans from various ethnic groups, as well as black scientists from Zambia, Zimbabwe, Lesotho, and the Democratic Republic of Congo. All had received their PhDs in South Africa—mostly at the University of the Witwatersrand (in Johannesburg) and the University of Cape Town, a few at the University of KwaZulu-Natal.

The bench scientists were a young group, often straight out of their PhD programs and generally less than a decade into their careers. Most were born in the 1980s, and so they were not quite "born free" (the term used to describe South Africans born after the transition to democracy, in 1994), but they came of age in an already-democratic context. Post-apartheid legacies informed the context into which they articulated their belonging, as scientists and as people. They were idealistic young adults, inspired by calls to contribute to "African solutions for African problems." In this chapter, I will first explore the bench scientists' articulations of their project and will then conclude the chapter with a consideration of how these scientists might exemplify the rainbow nation at the bench as they aspire to make the place from which they work a site of global science.

All the bench scientists at iThemba agreed to one-on-one interviews, which took place in a small conference room off the lobby in the front of the building and were squeezed in amid the flow of their activities at their computers and lab benches. I also conducted multiple follow-up interviews with many of them over the years. The small kitchen area / break room tucked between the reception area and the work areas provided opportunities for more sustained, informal conversations with those scientists who were most interested in engaging with me. It was an extraordinary privilege for me that the scientists were allowed and encouraged to spend significant time with me, on company time.

iThemba's organic-synthesis methods were indistinguishable from what might be done in well-equipped labs anywhere else in the world, and the work was informed by a network of advisers who were global experts. Whereas elsewhere on the continent laboratory scientists express anxiety that the medical and scientific work in which they engage is "not modern,"[2] at iThemba the laboratories were technologically up-to-date. For many observers, this participation in practices that conform with global standards might make iThemba seem "not African," in the sense that it did not conform to dominant narratives of Africanness. The lab was discontinuous with the stories of lack, distrust, and making-do that Nolwazi Mkhwanazi has suggested characterize the problematic single story generally told about Africa by medical anthropology.[3]

iThemba Pharmaceuticals' identity as a South/African company was not simple, but it was important. As we have seen in previous chapters, with its elite international board of cofounders and scientific advisers, iThemba was very much a global project as well as a South African one. In this chapter, I will leave aside the perspectives of the members of its global networks to focus on how the on-site scientists articulated their project. I will be particularly attentive to how the scientists articulate iThemba's work as rooted in place. In interviews with scientists at iThemba, local and global aspirations emerged together. The scientists talked about how their motivation to do this work came from personal experience with disease, ideas of democratic citizenship, and the opportunity to have a job "at home." These motivations were intertwined—and were all connected to place.

References to both personal experiences and the needs of the people are ways that the scientists at iThemba can be understood to be producing "situated knowledge." As feminist theorist Donna Haraway has argued, "the standpoints of the subjugated are not 'innocent' positions. On the contrary, they are preferred because in principle they are least likely to allow denial of the critical and interpretive core of all knowledge."[4] Elite Africans like these scientists are not precisely "subjugated"—I'll return to their betwixt-and-between status below—but their vantage points from a peripheral space of global science did become a location of insight. Like all knowledge producers, the scientists at iThemba were looking *from* somewhere; in contrast to most Global North scientists, they were highly aware of that.

Unlike the management and Scientific Advisory Board of iThemba, the local bench scientists were not public figures, and so I try to make sure that individuals are not identifiable when I quote them below, even though this has the unfortunate effect of making their voices somewhat disembodied. I include longer block quotes in an effort to preserve more sense of voice, and the articulations are rich.

Rooted in Personal Experience

One way that iThemba's work was rooted in place is connected to the scientists' personal experience with disease. The company was founded with a focus on HIV, TB, and malaria, and some of the scientists were drawn to drug discovery science by their own personal experiences with these diseases. For example, I asked one scientist who was from a country neighboring South Africa to the north, "So, did you always want to be a scientist? Even when you were a child?" The scientist responded, "Yes, yes. Actually, my interest in drugs started when I was about fifteen or sixteen. I got struck by malaria the first time ever. And I was so sick I thought I was going to die. I had never been so sick in my life. So I thought, after I recovered, I thought I'm going to make a difference helping people." There may be scientists in the Global North who also find their motivation in embodied experience of disease, albeit not likely malaria. But there is a specificity to these scientists' tie to the locally relevant diseases that formed the company's mission. At the same time, choosing embodied experience as the starting point for the path to science speaks to a distinctive integration of labor, intellect, and care that feminist sociologist of science Hilary Rose calls "hand, brain, and heart."[5]

The scientists' experiences were not only personally embodied but also informed by social context. More frequently, the scientists at iThemba spoke not of their own physical experiences with disease but of intimacy with the diseases in their communities and their region as being inextricable from their sense of urgency in doing the research. For them, these diseases are not merely abstract "matters of concern," in a Latourian sense, but also "matters of care," in which they are hailed not only to reflect upon but to act.[6] As one scientist put it:

> I think that it is important that people who are being affected are the ones who are doing the research. Because if you have seen someone suffer, you make all the effort to ensure that whatever you are doing gets out and ordinary people can benefit from it. You clearly understand the importance of doing it. Even people from other countries, they still understand, but the fact that it has affected you, you feel the strong need of intervention, the strong need of finding something that will be helpful and will be accessed in a cheaper way.

The close tie between research and access that this scientist articulates here contrasts with the conventional framing in science and technology policy, as described by policy scholar Susan Cozzens, in which the utilitarian engine of innovation and the distribution of its products are considered separately, and "any unequal distribution of benefits is an unintended consequence of policies that are fundamentally civic-minded and good."[7] For this scientist, it is human ties with those suffering from the diseases of inquiry that make the issues of innovation and access inextricable. I asked a follow-up question: "And have you felt that impact of malaria, TB, and HIV?" The scientist responded, "Not malaria, but HIV, yes, absolutely. There are many people who have died of HIV and TB who are very close to us. Especially HIV. They always say that if you are not infected, you are affected." It is certainly true that people around the world are impacted by HIV, and individual scientists anywhere may find research motivation in personal experience. Yet in South Africa, the sheer scale of HIV creates a radically different terrain. Of the approximately 35 million people living with HIV in the world, an estimated 18 percent live in South Africa.[8] The prevalence of HIV infection in South Africa is among the highest in the world, estimated to be 18.8 percent of those aged fifteen to forty-nine.[9] Although HIV/AIDS is no longer the leading cause of death in South Africa, as it was in the early 2000s when these scientists were coming of age, it remains among the top few causes. As of 2013, HIV was the third leading cause/ of death in South Africa, after tuberculosis and influenza/pneumonia.[10] This global disease is very meaningfully local.

HIV informs South African subjectivity more broadly, including in the post-apartheid AIDS activism that I discussed in the previous chapter. As anthropologist Steven Robins has argued, for people living with HIV in South Africa, the highly impactful Treatment Action Campaign and other social movements have provided a route toward becoming simultaneously political activists and responsibilized citizens.[11] This is the context in which the iThemba scientists tried to do their part—not outside standard liberal rhetorics of individual responsibility but not *only* in those terms. The Treatment Action Campaign rejected the cultural nationalism embraced by HIV denialists—notably Thabo Mbeki—and instead "created the political space for the articulation of radical forms of 'health citizenship' linked to a genuinely progressive project of democratizing science in post-apartheid South Africa."[12] The scientists at iThemba didn't identify as political activists, and the term "responsibilized citizens" would have been alien to them, but this concept resonates with their political consciousness and their sense of obligation. The scientists at iThemba extended the responsibility for realizing effective and accessible treatment for South Africans living with HIV beyond where that responsibility is usually placed—on HIV-positive individuals, [13] Big Pharma,[14] and the South African state[15]—taking a share on their own shoulders as drug discovery scientists.

Rooted in Democratic Citizenship

This sense that all southern Africans are affected by HIV is connected to a second way that the iThemba project was articulated as rooted in place. For the racially diverse scientists involved, the importance of "taking responsibility for AIDS" and other infectious diseases was often linked with the moral imperative for South African scientists to focus on the needs not only of the global or South African elite but also of the poor and black South African majority, as well as to spread that benefit to other countries on the continent.

To the extent that iThemba was successful in making compelling claims on both South African publics and global science, the African scientists became, like the African American geneticists described by sociologist Alondra Nelson, "bio-culture brokers": "they deploy the 'authentic expertise' . . . available to them as both minorities *and* community-minded professionals to forge consensus in and between biomedicine, scientific domains, policy circles and the public."[16] They positioned themselves as native speakers of two discourses, global science and African democracy, uniquely situated to make the former accountable to the latter.

The locality of the problems was not understood as an issue of racialized bodies, as it can be in arguments for local pharmaceutical research elsewhere, notably Japan.[17] Although genetic notions of belonging can have salience in South Africa,[18] the belonging here was not so molecular.[19] It meant being accountable to democratic society. One scientist, a black South African who grew up poor, told me:

> I think also we have to do like the first world countries. When you go to the United States or the United Kingdom, they have high investment in research and development. And research in the end of the day works for them. Versus in Africa—of course we have highly talented people in this country and in Africa in general, but we haven't seen much of science responding to the actual needs. What it means for science to address the needs down in my home town, where they are facing [TB and HIV]. And so that is where my social aspect comes from, is how do you make science respond? Science and education need to address what we are facing in this country.

This scientist's articulation of northern science's responsiveness to the needs of northern people is too generous an assessment, but the projection does important work. It locates commitment to local needs within the most advanced science rather than, say, as an outside force that corrupts science.[20]

Another scientist articulated the obligation to solve African problems in a complementary way:

> We conduct research in neglected diseases such as malaria, tuberculosis, HIV and AIDS. These three diseases affect South Africa more—well, apart from malaria—TB and HIV/AIDS affect South Africa more than any other country in Africa. . . . And unfortunately, the big companies, over time, didn't invest any money into discovering new therapies for TB. So I think over a period of forty years there haven't been any new drugs for TB. And the onus is on companies like iThemba, located in South Africa, where the majority of the people are suffering, where a majority of cases of TB are located, to make a difference. And we'll make that difference by putting our ideas on paper, taking them out of the paper, into the lab, and making something that we can develop over time into a TB drug or HIV drug.

The "onus" of the local need to address TB and HIV becomes both an opportunity and an obligation.

Another scientist, after having been trained in natural-products synthesis, foregrounded a broader desire to be useful and enjoyed the shift to solving real problems:

> If I am making this molecule, it's to test; it's not just to make for the sake of making. It's much more rewarding. And I always wanted to get into something application based, and especially with a medicinal aspect. It's quite important to me to have some sort of real purpose behind it. I don't think I could find full satisfaction making, I don't know, flavors. It's different. It's got a very direct need, and you're working towards solving real problems.

The desire to solve "real problems" certainly drives many practically minded scientists around the world—including many working in drug discovery. But for this scientist, the "real problems" were also located. The problems being faced by people in South Africa became a source of motivation, in a way that was simultaneously overarching and mundane:

> Well, why it's important to me is that it means I can still stay here. But it is, there's just a feeling, that it would help move the country forward. And we are the ones who can see the people dying. We know the people who have HIV. Know people, you know, who have malaria. We are very much in the thick of it, so we can, we know that there's pressures given us. It's not, sometimes, not often, but sometimes I go home and I think—oh, I'm tired, I just want to go home, and then you think of the side of it where we are meant to be working on this cure, and how can we possibly slack off for a moment? The need is so obvious. And most of the time we forget about the need and we just focus on the molecules and don't really think about the end goal because the end goal is so far away. Every now and then, you remember why we are here and why we are doing it, and how important it is.

In this articulation, being located "in the thick of it" fostered a sense of urgency. When the focus was zoomed in to the molecules, as it often was in day-to-day work, context could slip away. But that slip was temporary, and being located in South Africa meant that the scientists were constantly re-presented with the real need.

An explicit allegiance to Africa and South Africa was common, and sometimes very quid pro quo. For example, one scientist described the obligation to continue in science because of the investment that the government had made in their training:

> When I looked at the options, iThemba was really one of the only places where I could do what I trained to do. And it seemed a waste actually, if you go up and just get a job in a different discipline. We are highly specialized, and I feel a bit of a social obligation, because I have had scholarships since I started studying. And you think about how much money the government has pumped into you, and I feel it's my social obligation to actually use it in the field I was trained in, because that's why they are putting the money in there. They don't, they aren't holding me to that at all. But I want to give back, to give back into chemistry.

Two intertwined citizenship obligations are being articulated here: giving back to "South Africa" and giving back to "chemistry." iThemba provided a route to fulfilling both.

There were occasional exceptions. One scientist described frustration with the company's focus on HIV, TB, and malaria and lamented that "these diseases are neglected for a reason." From this perspective, South African scientists should share the same focus on diseases like cancer and Alzheimer's that global pharmaceutical companies overwhelmingly have, because that is where the most scientific and market potential is located.

Yet for most of the scientists most of the time, context and project choice were inextricably linked. I asked one scientist, "Besides your own interest in doing the work, do you think that there's value in drug discovery taking place in South Africa? Or does it not really matter where drug discovery happens?" The reply was multilayered:

> It doesn't, to me it doesn't really matter whether the cure comes out of another country or here, or whether the drugs come out of another country or here. But obviously it's important for the South African economy in the broader sense. If we are seen to be producing our own drugs for our own diseases, it will increase our standing in international science. I think it will make the South African people feel less helpless towards these diseases. If they know that South Africa is providing something for them, we are looking after ourselves. So, kind of, I guess, it's a yes-and-no answer. I think the difference it will make is if the drugs come out of South Africa, there's going to be a difference in the sense people have of South Africa. It's not going to be a concrete difference of it works or doesn't it, because obviously a drug is a

> drug. But it should help us produce them more cheaply. It should help build international confidence in South Africa, and it should help even the people on the street to feel looked after by their country. Not to feel like a third world country relying on the rich countries of the world.

On the one hand, this scientist categorically rejects the salience of place: "a drug is a drug." On the other hand, there are layered invocations of location: "increase[ing] standing in international science," "looking after ourselves," and "help[ing] people on the street to feel looked after by their country." Multiple political articulations are at stake here.

We might read these articulations as questions of representation: iThemba's project would change how South Africa is represented to its own citizens and to global observers. As postcolonial theorist Gayatri Spivak has canonically argued, "representation" carries a dual meaning: "representation as 'speaking for,' as in politics, and representation as 're-presentation,' as in art or philosophy."[21] This blurring between acting as a proxy and depicting has been important in science and technology studies, both classic actor network theory by Michel Callon[22] and specifically postcolonial science and technology studies by Cori Hayden.[23] The scientist quoted above is speaking for South African patients and trying to change the image of South Africa both for its own people and for global observers. Scientists are by definition elite, and so they are not simple representatives of the broad population of southern Africans affected by TB, HIV, and malaria, and yet their sense of obligation to faithfully reflect that larger population's interests and speak for them on an international stage points to ways in which they are connected.

The scientists, especially those from non-elite backgrounds, were constantly translating between scientific and lay languages. Those without scientific literacy were particularly valued. Asked how they describe iThemba to a lay audience, one scientist told me:

> The first thing is to boil down to make an example. You say we are trying to make things that have to do with pills. They understand tablets, they understand that better. So for us, we are more into understanding what's inside that pill, and what does it do to make you feel better. If you have a headache, you take a Panadol [a brand of acetaminophen]—you feel better. So what I am interested in is what is in there that makes you feel better, and how does it work inside you.

Who is being spoken to influences how the work is described:

> If they have an academic, some scientific background, I can say, "Look, we are looking at chemistry, and how you can make chemical compounds that respond to different processes that are being affected in these diseases. But at the end of the day we are actually looking for new active ingredients to provide a cure." It's a different vocabulary. You have to, really. That's the situation in our country; you have to really try adjust it to different levels of people that you come across. You can't assume that people are going to understand what we do; you have to use words they can understand.

For most of the scientists, their articulation of their mission always returned to the desperately poor. Many examples came up in the interviews, with frequent articulations of HIV and TB as diseases of poverty, wherein one barrier to access to treatment—even if the pills were free—would be lack of access to food to eat with the pills.[24] One part of the situatedness of these scientists is that they were all aware that HIV as a medical problem is inextricable from HIV as a social problem, and that biomedical solutions are just one part of the need for social conditions compatible with health. As feminist philosopher Sandra Harding has pointed out, "from the perspective of Northern labs" it can be hard to see that "Northern residents, men or women, will probably not be the most valuable agents of democratic social change in science and technology worldwide."[25] From the perspective of this South African lab, it is evident that African scientists and citizens will have to lead if they want change to happen.

The social awareness on the part of the South African scientists often struck me as resonant with that of African American physician-researchers in US-based research projects.[26] Relative to those from dominant communities, African American physician-researchers and their professional organizations recognize that both biomedical and social interventions are essential to improving the health of marginalized communities within the United States. Contrary to much of the literature about pharmaceuticalization,[27] for many of those physician-researchers, pharmaceutical interventions are not an alternative to social and political interventions but a high-stakes component of it. There was a parallel recognition on the part of iThemba scientists that scientific innovation and social change are inextricable and urgent.

Even though none were activists, the scientists' participation in their society and their participation in their field were intertwined. As Sandra Harding has argued, on a more abstract plane, "Interests in particular kinds of scientific and technological projects are created as part of social transformations, while such social transformations are shaped and directed in part by the kinds of scientific and technological projects that can be put in service to their goals, explicit or implicit. To put this point another way, feminism and postcolonialism both argue in effect that how we live together both enables and limits what we can know, and vice versa."[28] Where these scientists lived affected their research agenda, in a way that fueled rather than detracted from their science.

The conscious awareness of the fact that scientific research is done in a social context permeated the sensibilities of the scientists working in this South African lab. This was particularly striking to me because my first visit to iThemba came shortly after my first conversation with iThemba's Atlanta-based cofounder Dennis Liotta, who described his work leading up to second-generation antiretrovirals in a highly performative way. He said that when he first started doing HIV-related research, he was "so naïve." He said that he had thought that "if just we could find the chemical that would cure AIDS, it would be solved." He and his collaborators found drugs that treated AIDS effectively over a long term, but did not cure it. Other aspects of the problem, however, came as a surprise to him: "If the stigma was too great, people would not get tested, so they would not get the drug. Or if the cost was too high, they wouldn't get it." It was only over time that he came to see what the South African bench scientists could not help but know: that AIDS as a biomedical disease is inseparable from its social and political context.

Rooted in Working "at Home"

The third way that iThemba was rooted in place is that it was seen as a way to do meaningful work "at home." iThemba was working to make medicines *for* the people—low-cost medicines for neglected diseases. But it was also working to make medicines *by* the people. This touches again on the dual sense of representation—as depiction and proxy. As a racially diverse group of highly educated people, there was a tension in whether the scientists at iThemba were themselves "the people" or whether they were "serving the people." They now were part of the multiracial middle class that inhabits gated communities and other elite spaces north of Johannesburg's central business district, not the many spaces of ongoing poverty south or outside the city (or the nearby township of Alexandra).[29] The scientists were themselves part of the southern African people, albeit a relatively privileged part, and they were trying to help themselves too. Several scientists answered my question about why it matters that drug discovery happen in South Africa with variations of "Well, it means that I can stay here," in a familiar place close to family and friends.[30]

I asked one scientist, "Is it important that R&D work happen here in South Africa?" The scientist responded:

> I think it's been long overdue. The country has produced a lot of top-class scientists, and it's only fair that they should be a lot more involved in research and development. Obviously, it puts the country on the global map. But you also want to create employment opportunities for a lot of the students that are studying chemistry. One of the major questions that science students have is, Once finished with these studies, what next?

Thus, if drug development were to take place in South Africa, it would be "for" chemists in South Africa as much as it would be for desperately poor patients coinfected with TB and HIV. Of the relatively small number of scientists trained in Africa, only a small portion both continue to work in science and stay in Africa. These scientists see their site of work as a commitment to a place, and a privilege.

If slogans such as "African solutions for African problems" can seem to prioritize the local over the global, that is only part of the story. The local is essential as the space of the work, and its context, but it is also a route to global relevance—putting "the country on the global map."

One scientist described observations of more senior scientists involved with the first iteration of iThemba, whom they had met during undergraduate studies at the University of Cape Town:

> And they were on the same floor that I was working on. So they were at the university. And I remember that what they did was secret. But they were very helpful. So whenever anything went wrong, you could go knock on their door and both of them would help me and they always seemed to be having fun. So the first impression was, wow, this looks like a lot of fun what they are doing there. And obviously then we heard that they had run out of money and it was gone. That was a bit disappointing because you start wondering, where I am going to land up working? Can I actually do synthetic chemistry in this country or am I going to have to look in other countries? So when rumors came that they were going to start again, it was very exciting.

By the time that the scientist was pursuing a PhD at the University of the Witwatersrand (Wits) in Johannesburg, the head of the relaunched iThemba was spending quite a bit of time there:

> And [the former head of iThemba] spent a lot of time at Wits meeting with students and speaking to them, and so we had a lot of feedback on what was going on. And he came and met us a bunch of times. So it was always, there was this question in the background, is iThemba coming back, are we going to have somewhere to go? And it was, if they can get back up and running I want to work there. Which it was my dream job, so to actually get it was very, very exciting.

The bench scientists had a high degree of identification with the goals of the company. That meant that they had to manage their own hopes, aware that the ultimate goal may never be reached: "And there's a little bit of a sense of there's not all that much chance that any of our molecules will actually ever making it to drug. Which, of course, is the reality of being at this end of the chain, unfortunately." As part of iThemba's business model, scientists also worked on contract chemistry projects, synthesizing molecules for other companies on contract in order to provide cash flow for the company. Working on contract chemistry projects thus had an ambivalent aspect—on the one hand, less meaningful and a distraction from drug discovery; on the other, more assuredly productive and contributing to the company's sustainability:

> But I'd rather be working towards actually making something that can make a difference than working so that we can get funding so that we can get paid for a certain job. That, there seems to be a slightly negative spin on the whole thing in that—let's just secure the next project, because then they'll be able to contract those, and we'll carry on doing stuff, to occupy the time, and I'm really hoping that we can move forward and get stuff out there in whatever way we can.
>
> At the moment I am just working on contracts this week. Which I don't mind, because although it's not curing malaria or anything, I am making something we can sell, which can make some money, which will help the company continue to grow. I don't mind doing the contract work, because I obviously want this company to be sustainable, and in order to do that we need to produce some of these products. So hopefully a drug eventually.

The scientists' high level of identification with the company marks them as elites. More broadly, since the scientists at iThemba were now middle-class, educated individuals, all living outside the malaria zone, their commitment to science in the service of the people exemplifies a "split and contradictory self."[31] If we understand their project as situated knowledge, we can see that "positioning implies responsibility for [their] enabling practices."[32] The political is also the personal: trying to find a way to make a living and a life in real time.

The imperative to succeed comes not only from outside but also from within. As one bench scientist put it:

> I think, well, obviously being based in Africa, also, more people are going to support the company and put in funding because we are Africans working for African disease research. It makes a difference. Obviously, being the, well, let's call ourselves the only pharmaceutical company . . . the government is very interested in fully supporting us, which is a very meaningful thing to have behind us. And I don't know, being based in South Africa, I think, well, we've got a reason to try. We really, we need to make this company succeed, or we'll all be out of work. We'll have to go into different niches. It goes up to quite a lot of motivation and enthusiasm.

The scientists at iThemba were good examples of what Warwick Anderson calls "conjugated subjects." The project has little potential to inspire any "romantic vision of ethnoscience."[33] The knowledge was being made in a postcolonial context, but it was discovery not recovery, and it was not subjugated knowledge. Although this South African project might be understood to be nothing more than an outpost of hegemonic northern science, it became more than that as the scientists made it their own, by connecting their work with personal experience, democratic citizenship, and the privilege of working at home. Anderson compellingly argues that science and technology studies scholars should pay more attention to postcolonial hybridity; what Homi Bhabha might flag as their mimicry of colonial forms is patently hybrid.[34] To be anglicized is not the same thing as to be English, and although iThemba's project may not be subversive, neither is it merely derivative. The iThemba project sought to undo the tight coupling between northernness and the capacity to participate in global science.

The Rainbow Nation at the Bench

Strikingly, the bench scientists tended to use "African" and "South African" interchangeably in interviews, whether they were from South Africa or from further north. In this sense, the Africanness of iThemba's scientists can be understood to be aligned with anti-apartheid and post-apartheid discourses. For example, it is resonant with that of the Freedom Charter of 1955, a statement of core principles of the African National Congress and allied anti-apartheid organizations: "South Africa belongs to all who live in it, black and white."[35] Historian Saul Dubow quotes this opening line and argues that it imprinted itself on the consciousness of most South Africans only after the 1990 unbanning of the African National Congress and Communist Party—when these bench scientists were small children in South Africa and neighboring countries.[36] This notion was incorporated into the preamble of South Africa's 1996 Constitution, which declared that "South Africa belongs to all who live in it, united in our diversity."[37] The Constitution is a particularly important touchstone in South Africa. Jean Comaroff and John Comaroff point out, "In South Africa, the Constitution—both its content and the found object itself, as aura-infused in its little red-book reproductions as in the original—is *the* populist icon of nationhood."[38] And so this "nonracial" articulation of belonging matters.[39] As Saul Dubow has argued, "There is nothing self-evident about South Africa or South Africans and the struggle for South Africa has always been, and in many ways continues to be, a struggle to become South African."[40]

iThemba scientists' notion of "African" is also resonant with Thabo Mbeki's famous "I Am an African" speech on the occasion of the passing of the 1996 Constitution, in which he avowed connection to both the South African landscape and the diverse populations living in South Africa.[41] He claimed that he owed his being to the indigenous Khoi and the San, as well as migrants from Europe, Malay slaves, specific Xhosa, Sotho, Zulu, Venda, and Tsonga leaders, and more distant peoples of the continent: Ethiopians, Ashanti of Ghana, Berbers of the desert. As feminist legal scholar Laura Foster has pointed out, "Mbeki's speech asserted racial difference by attending to South Africa's ancestral pasts but subsumes difference under a privileged universal Africanness. It simultaneously marshaled difference and sameness to craft new terms of belonging for the nation-state predicated upon a shared African identity."[42]

African identity becomes place based, albeit on the scale of a continent rather than a country. In Mandela's inaugural speech in 1994, he was more immediately local as he argued that "each one of us is as intimately attached to the soil of this beautiful country as are the famous jacaranda trees of Pretoria and the mimosa trees of the bushveld."[43] Mandela's choice of botanical examples is particularly striking in that these plants are not "native": jacarandas are originally from South America, mimosas from Australia. The democratic South Africa would pointedly include waves of immigrants, human and botanical, in the landscape. Geographical roots point to a unity of copresence, unlike a unity based on shared ancestry, and they provide the basis for Mandela's ambitious project: "We enter into a covenant that we shall build the society in which all South Africans, both black and white, will be able to walk tall, without any fear in their hearts, assured of their inalienable right to human dignity—a rainbow nation at peace with itself and the world."[44]

Even the famous moniker "rainbow nation," coined by important African National Congress figure Desmond Tutu and popularized by Nelson Mandela, points to a tension: the rainbow is recognizable by the presence of multiple distinct hues, and yet it also exists as a unified object. The rainbow also represents hope after the storm. Mandela continued, "We are both humbled and elevated by the honour and privilege that you, the people of South Africa, have bestowed on us, as the first President of a united, democratic, non-racial and non-sexist South Africa, to lead our country out of the valley of darkness."[45]

It's worth noting that the intertwined discourses of nonracialism and pan-Africanism are distinctly elite. Sociologist David Matsinhe has argued that "the Africanness of South Africa and the South Africanness of Africa remained a thing of the elite while the lower tiers of the figuration remained strongly parochial and chauvinistic."[46] The scientists involved with iThemba, not only on the advisory board but also on the bench, were certainly elites.

Like the declarations in the Freedom Charter and in Mbeki's and Mandela's speeches, iThemba's notion of Africanness embraced white settlers in South Africa. Like Mbeki's articulation, it also embraced those from across the continent. Yet there are differences too. Whereas Mbeki and Mandela hark back to the landscape and to past greatness, the bench scientists eschew that kind of mythologizing and poetry.

The post-apartheid cosmopolitan space of innovation that iThemba instantiated emerged in African contexts but not necessarily the ones most frequently associated with the continent. The Africanness of the project was not rooted in nature or in tradition. These scientists' endeavor was not African in any ethnoscientific sense. The referent "African" here meant the *people*, not the plants. Moreover, it meant African *scientists* rather than, for example, African *traditional healers*. Yet for the scientists, the work of iThemba was tied to place. Geography mattered, albeit not because of any special relationship with African nature.

For iThemba scientists, Africa was the place that their synthetic laboratory was situated; the continent was not itself a "living laboratory."[47] Any Africanness of the project was necessarily tied both to particular histories—including legacies of colonial resource extraction and apartheid that I have discussed in the previous two chapters—and to a modern, cosmopolitan present and future. In this cosmopolitanism, like that theorized by philosopher Kwame Anthony Appiah, commitments to local and regional communities, on the one hand, and to global communities of science, on the other, were aligned rather than in conflict.[48]

Spatializations of Pharmaceuticals' *Subjects* and *Objects* Should Be Understood Together

From early on in the epidemic, "African AIDS" has been figured as an elusive object of knowledge about which Western bioethicists and others could opine, oblivious to what Africans with expertise might have to say on the matter.[49] This chapter has focused on Africans with PhDs who are striving to create new global knowledge about HIV/AIDS as well as TB and malaria, attention to whom is long overdue. The place of knowledge-making subjects matters, not just the place of the distribution of technoscientific objects.

Paying attention to educated Africans who work indoors is an important corrective to the type of writing that Binyavanga Wainaina satirizes in his scathing essay "How to Write about Africa"; while so much writing about Africa relies on tropes of abject but romanticized Africans, writing about African scientists provides a route to more nuance.[50] There are limits to the degree to which the individual scientists have been portrayed as individual people, because of the constraints of the corporate ethnographic methods that led me to protect the identities of the scientists. Nevertheless, I hope that these scientists have emerged as subjects.

Separation of subjects and objects is a fundamental binary of colonialism, as canonically described in Edward Said's *Orientalism*: "because of Orientalism the Orient was not (and is not) a free subject of thought or action."[51] Troubling that binary was part of what was at stake in iThemba's postcolonial science project. This suggests that the spatializations of pharmaceuticals' *subjects* and *objects* should be understood together. In iThemba scientists' search for "African solutions for African problems," different regimes of value jostle, such that ethical value, economic value, and epistemological value comingle.[52] For these scientists, ownership of pharmaceutical intellectual property would not only make an economic claim and an epistemological one but also locate a moral claim. In this way, they stake out for themselves the role of economic, epistemological, and moral agents, not merely passive recipients.

Ambitious African scientists were striving to be the subjects of scientific progress, which is to say, practitioners of scientific research at a global level.

They did this through articulating what anthropologist James Ferguson describes as the "place-in-the-world" that Africa might occupy: they wanted to replace hegemonic framings of Africa as a place of exclusion "nearly synonymous with poverty and failure" with their conception of Africa as a site *both* of people with urgent needs *and* of capable problem solvers.[53] Place still mattered: they cited their colocation with important *objects* of medical knowledge—impoverished Africans with urgent unmet needs—to make a claim for a role for themselves in global networks of science. Their cosmopolitan aspirations were rooted in their context. It's notable that the hoped-for drugs were context specific without being expensive, and so they might be uniquely situated to be democratic.

A synthetic-chemistry laboratory is an unconventional place to ground an analysis of African knowledge making. As philosopher Valentin-Yves Mudimbe has argued, "Africanists—and among them anthropologists—have decided to separate the 'real' African from the westernized African and to rely strictly upon the first."[54] Scholarship in my own interdisciplinary field of science, technology, and society does not generally share this scholarly tendency to ignore the knowledge made by westernized Africans, though the disproportionate attention to efforts to translate ethnoscience into global science might follow from this tendency because the plants and traditional healers can stand in for "real" Africa. Laboratory scientists are undoubtedly "westernized."

These scientists were trying to make indigenous pharmaceuticals of a very particular kind: not autochthonous but meaningfully their own. This represents an intervention into the framing described by Clapperton Chakanetsa Mavhunga: "science, technology, and innovation seemed to be things inbound from somewhere outside Africa," and "the basis of the conversation about Africa was that it was the recipient of science, technology, and innovation, not a maker of them."[55] (For Mavhunga, bench science cannot be a useful corrective, because it remains elitist, though iThemba scientists would beg to differ.)[56] Along similar lines, Gabrielle Hecht's topical focus, nuclear technology, is very different from pharmaceutical technology, and yet there are important resonances. She writes: "Technology's absence from analyses of African political agency . . . makes it appear exogenous—a global force that buffets Africans and turns them into victims. Such a view makes it difficult to grasp how technological entanglements permeate industrial labor in postcolonial Africa, how these entanglements both open and close political possibilities, and how their contradictions sometimes serve as sources of hope."[57] If pharmaceutical knowledge making is figured as external to Africa, that positions Africans as victims of its global winds. What might it mean to locate hope in African-discovered pharmaceutical technology?

In slogans about "African solutions for African problems," Africa is invoked in complicated ways. According to postcolonial theorist Achille Mbembe, "whether produced by outsiders or by indigenous people . . . discourses on the continent are not necessarily applicable to their object. Their nature, their stakes, and their functions are situated elsewhere. They are deployed only by replacing this object, creating it, erasing it, decomposing and multiplying it. Thus there is no description of Africa that does not involve destructive and mendacious functions." And yet that does not render such understandings of Africa inauthentic: "But this oscillation between the real and imaginary, the imaginary realized and the real imagined, does not take place solely in writing. This interweaving also takes place in life."[58] For these iThemba scientists, Africa is inescapably and simultaneously real and imaginary.

iThemba's project was aspirational more than actual, fittingly for a company whose name was the Zulu word for "hope." As Sheila Jasanoff and Sang-Hyun Kim have argued, imagined possibilities matter; what they describe as "sociotechnical imaginaries" are vital sites for thinking through not just what technology should be but also what the nation should be, because "technoscientific imaginaries are simultaneously also 'social imaginaries,' encoding collective visions of the good society."[59] The idea of "African solutions for African problems" captures a sociotechnical imaginary organized around locally discovered, innovative drugs for infectious diseases that articulates South African medical research and society in ways distinct from a range of alternative imaginaries—of blockbuster drugs, aid complexes, plant-based therapies, or pharmacogenomics. It is an imaginary that is particularly useful for thinking through postcolonial technoscience circulation in the Global South beyond diffusion and translation.[60]

FIVE

Im/materiality of Pharmaceutical Knowledge Making

The workday at iThemba Pharmaceuticals started at 8 a.m., as is typical in Johannesburg, but I usually arrived later—at about 9:30, in time to get settled in for the morning tea break at 10 a.m. Over the period of my research there, which involved several trips between 2010 and 2015, the 10 a.m. tea break was one marker of the ebbs and flows of the morale among the bench scientists at the small pharmaceutical company—whether they spent this time in the kitchen making small talk or whether they spent it at their computers.

The flow between computers and labs marked a key part of the rhythm of the workplace. Typically, when I arrived in the mornings, most of the bench scientists were in the computer room situated at the center of the building, searching the published scientific literature for information relevant to the chemical reactions that they were working on. This schedule was motivated, in part, by time zone: the scientists sought to access the databases before the United States woke up and the databases slowed down.[1] Many other routine tasks, ranging from ordering chemical reagents to writing reports, also took place at the computers. From morning tea onward, the scientists would gravitate toward the labs, spacious rooms that extended along both sides of the building. By the afternoon of a typical day, most of the scientists had donned their lab coats and were performing reactions and analyzing molecules. I would come to see the scientists' circulation between computers and labs, between working with digital documents and manipulating material stuff, as an instantiation of the im/materiality of pharmaceutical knowledge making.

In the previous chapter, we saw that "Africa" is simultaneously a physical place and an elusive idea. When the diverse young bench scientists working at iThemba were inspired by the slogan "African solutions for African problems," the "Africa" that they invoked was simultaneously real and imaginary.[2] Here, I will explore ways in which pharmaceutical knowledge too has evocative material/semiotic ambivalence. Pharmaceuticals encapsulate and carry both matter and meaning.[3] Pharmaceuticals' materials are heterogeneous, and their capacity for meaning making is both informational and symbolic. For pharmaceutical information to become drugs, it must be materialized with ingredients and processes that are unevenly distributed in space.

This chapter explores wide-ranging ways in which information and materiality were entangled in efforts to create drug discovery capacity in South Africa: the material logics of chemistry; the symbolic, literal, and virtual "place on the map"; aspirations for a "knowledge economy"; and patents and pills. I highlight contrasts with information technologies as I explore potential pharmaceuticals and the making of them as fascinating material-semiotic objects and practices.[4] Drawing together disparate threads from iThemba scientists' perspectives, philosophy of chemistry, and theories of the digital, I argue that attending to these plural, simultaneously material and informational registers shows the allure and the difficulty of materializing the intellectual property (IP) of pharmaceuticals.

Making Chemical Compounds, Making Pharmaceutical Knowledge

In the context of a drug discovery lab, the material and informational qualities of chemicals are inseparable. As anthropologist Cori Hayden has pointed out, "Chemical compounds are forms of materialised scientific knowledge that also happen to be the products of literal practices of distillation, isolation, and reduction."[5] This is precisely the kind of simultaneously abstract and material work that the scientists at iThemba were engaged in as they moved back and forth between the computer room and the labs. Attending to the perspectives of the bench scientists who were working at iThemba can open up considerations of materiality that resonate with anthropology of pharmaceuticals, philosophy of chemistry, and feminist new materialisms.

At iThemba, where the scientists were making laboratory-scale chemicals rather than industrial-scale ones, the necessary materiality of chemistry was intimately experienced—and bench scientists like working with interesting material things. Because of the scarcity of science jobs in South Africa, many chemists wind up leaving the lab and working in analytical fields such as banking, and the bench scientists at iThemba felt privileged to be able to work with materials with interesting emergent properties. As one pointed out in an interview, "And I love it, I love the lab work. So getting away from report writing, to actually get out and mix things together. And the fizzes and pops and colors and bangs and flames and—[*laughs*] it's exciting." I remarked that this scientist was the one who, in a lab meeting earlier that week, had mentioned liking something that exploded, and they responded, "Yes. I mean, it's not that I'm not scared. I'm always a little bit apprehensive of new, scary chemicals and treat them with due respect, but it is fun to work with exciting substances." These comments highlight both the wonder of chemistry and the need to pay attention to safety, which is a concern that has shaped the space of chemistry labs since their inception.[6] I was struck by the childlike joy that this generally quite sober PhD-level scientist was expressing here, and I also want to call attention to the word "respect." The scientist's mode is playful but serious, and the relationship with the materials is not domination or submission.

The affection for the materials that the scientist is describing is not exactly empathy. Its relationality is distinct from what feminist scientist Evelyn Fox Keller has characterized—following biologist Barbara McClintock—as "a feeling for the organism." That mode of relation requires having a feeling for individual organisms (in McClintock's case, plants) as they interact with their environment, in a way that is analogous with the way that human organisms interact with their environment.[7] In contrast, since the synthetic chemist starts with purified substances rather than complex living things, the interactions that the chemist incites and observes are at once too isolated and too partial to empathize with in this way. The synthetic chemist's relationality is also distinct from the blurring of boundaries between human and nonhuman in primatology and entomology that can allow reconceptualization of human social orders.[8] The relations among chemicals are too alien to provide a ready analogue for relations among people. The scientist's relationality is also different from the kind of oneness with the material that many craft workers, for example, describe, in which worker and material become one: the synthetic chemist's relationship with material is not colonizing or totalizing. The recognition of the fundamental alterity of the chemicals is preserved.

This scientist's perspective instantiates insights from the history and philosophy of chemistry. In their history of chemistry, Bernadette Bensaude-Vincent and Isabelle Stengers have argued, "Linking the destinies of an individual and a molecule, chemistry defines very specific relations between man and matter: neither domination nor submission, but perpetual negotiation—through alliances or hand-to-hand struggles—among individual materials and human demands."[9] These linked destinies of individual and molecule were palpable at iThemba.

The work of synthetic chemistry is a way of engaging with materials that is both creative and careful, both open-ended and disciplined. In her fascinating piece "Philosophy of Chemistry or Philosophy with Chemistry?," Bensaude-Vincent argues that the reason for a philosophy of chemistry is because "over the course of many centuries chemists have developed a special access to nature and a special way of investigating and dealing with material substances, they have confronted a number of epistemological and ontological issues that are worth discussing."[10] Notably, "[e]mergence in the case of chemistry does not convey the presence of a mysterious vital force."[11] I find this to be a useful corrective to feminist and other materialisms that evoke a nostalgia for vitalism.

As discussed in earlier chapters, science and technology studies scholarship on drug discovery in Africa has overwhelmingly focused on bioprospecting and traditional knowledges, endeavors that can be hard to extricate from colonial romanticizing tropes. In synthetic-chemistry-based drug discovery, there is no originary purity of knowledge to be salvaged and recovered. Since the term "synthetic" is often used as an antonym for "natural," synthetic chemistry is a fitting object for a cyborg feminism that seeks to examine naturecultural worlds.[12] Donna Haraway's cyborg was famously not born in the Garden of Eden and does not desire a return to a state of innocence after some fall from grace, and the same is true for synthetic chemistry. Emergence in synthetic chemistry resists romanticization.

This does not mean that chemicals aren't plenty lively. As Bensaude-Vincent puts it, "chemical substances are not samples of a Cartesian passive matter."[13] That's in contrast to, say, the photons that are pushed around in physics. Indeed, chemicals' activity is distinctly interesting: "They are active and reactive individuals whose behavior is partly determined by the neighboring substances. Therefore they require a relational ontology."[14] In a profound way, you don't know what a chemical is until you know what it is next to. Consider: if you add either hydrogen or oxygen to a fire, you increase the fire, but you can put the fire out with the same chemicals bonded together in H_2O.

Whereas physics or biology can sometimes seem to describe a superhuman world, the way that some kind of capital *N* Nature "really is," chemicals exist in useful relationship to human and nonhuman infrastructures. "Once analyzed, purified, and characterized, chemicals are hybrid products of nature, instruments, and operations. Chemical molecules . . . exist as composites of nature and society, of theoretical potentials, social or economic pressures, as well as environmental requirements."[15] In other words, synthetic chemistry is necessarily naturecultural and sociotechnical in character.

Returning to the space of iThemba bears this out. Within the lab work, there was another rhythm, between mixing, isolating, and analyzing. Powdered reagents and liquid solvents would be combined in beakers for mixing. This was the fizzing and popping step that the scientist was excited by. After that, the scientist needed to purify the products of the reactions: the new molecules created would be isolated through means such as evaporation, and the resulting powders would be put into small vials (fig. 6). These vials could be placed into machines situated in the center of the lab spaces for analysis, to verify whether the desired molecules of the requisite purity had been created. These machines—including a mass spectrometer and a nuclear magnetic resonance (NMR) machine—analyzed the molecular structure of the matter that the scientists produced, which provided an assessment of the success or failure of the creation of desired molecules. Those vials full of materials that achieved the requisite characteristics might then be put to use in another process or shipped out to another company that had ordered them on contract. At the analysis stage, there was real capacity for frustration, satisfaction, even joy.

However, direct engagement with chemicals is only one part of the job of a drug discovery scientist. Engaging with documents and digital information in literature searches is also an inextricable part of the work of bench scientists in drug discovery. The same scientist quoted above explained, "So I'm meant to be spending 70 percent of my time on the bench doing reactions and reducing these compounds, but you land up spending a lot

6. Materials and equipment at iThemba. Photographs by Katherine Behar.

of the day doing research around the reactions. So, reading the literature, planning, preparing your reactions, rather than actually getting into the lab and doing them." In the flow of the lab work, the reading and the bench work are iterative: "We are wondering whether one of the molecules which is a hit against TB, where it's not yet working against the enzyme we think it's working against, so I am doing research around another enzyme to see if it is possibly interacting with that one." This speculation about possible interactions derives from the emergent character of chemicals as both information and matter. And the stakes of the work are underscored by the scientist's reference to TB and enzymes: the ultimate goal is to translate information into an intervention on flesh.[16] Chemicals are relational and emergent even before they interact with human bodies, and their relational and emergent elements transform again within them.

The scientist's comments about the circulation between the computer room and the lab bench and the interactions among chemicals connect with broader theories of pharmaceuticals as opportunities to think about matter and information. Philosopher and geographical theorist Andrew Barry argues in his important essay "Pharmaceutical Matters: The Invention of Informed Materials" that "molecules should not be viewed as discrete objects, but as constituted in their relations to complex informational and material environments."[17] The need for scientists to move iteratively between the literature and the lab exemplifies Barry's argument that "specific molecules exist in the informational and material environment of the laboratory. But they also exist in a legal and economic environment of other molecules developed by other companies."[18] The molecules that iThemba scientists created existed in necessary relation with those produced by other companies. Indeed, because iThemba scientists engaged both in purchasing intermediates for pharmaceutical research and in making intermediates on contract for other companies, this constant traffic in matter and information within and beyond the lab was inescapably present.

Drugs are made through chemical processes, and chemistry is highly and necessarily material. As Bensaude-Vincent and Stengers have pointed out, "Working with materials whose properties determined the methods and strategies of production, the chemical industry distinguished itself from other industries by its prominent material logic."[19] They argue that chemistry's lower prestige, relative to fields like physics, is related to both its tight ties with industry and its necessary materiality. Chemistry is too practical and applied to bear the prestige that physics does. And for overlapping reasons, synthetic chemistry is not an ideal route to the global esteem of pure science promised by, say, South Africa's "prestige projects" such as the Square Kilometre Array telescope, which has the potential to contribute to astronomical research on a globally unprecedented scale.[20] In their aptly titled *Chemistry: The Impure Science*, Bensaude-Vincent and her coauthor, Jonathan Simon, argue that "chemistry serves as the archetypal techno-science, unable to restrict itself to pure theory, but always involved in productive practice."[21]

This potential for productivity is part of why, despite its lower prestige relative to physics, participating in chemical knowledge making offers an excitement of its own. Bensaude-Vincent and Stengers note that "chemists have continually been forced to defend the specific autonomy and rationality of their science, because their concepts and their methods formed notes or crossroads among heterogeneous areas on the map of knowledge and because they held strategic but disputed places on that map."[22] Finding a place on that map holds considerable allure.

"On the Map"—Symbolically and Literally

In iThemba's work, the importance of space and distance was both palpable and reconfigured. In interviews, references to the "map" came up all the time. This map was partly metaphoric: iThemba provides "an opportunity to put the country on the research map." And yet the map was also importantly literal.

South Africa's simultaneous distance from and proximity to Europe shape the capacity for research. On the one hand, South Africa shares a time zone with Europe, and the ability of European scientists to speak to South African scientists during the day was framed as a significant competitive advantage for contract chemistry work and collaboration. As one bench scientist put it in an interview, "Our location is such that we can have collaboration between Europe and Africa. We don't have much time difference, so if we need to achieve deadlines on projects, we can coordinate very easily with people in Europe. And I think funding, we are trying to get funding for projects from Europe. So providing updates is much easier. I don't have to wait to come in tomorrow to get a response from the US. Or come in at five o'clock to check the response before the guys stop working." This capacity for virtual connection also collapsed some distance between north and south with regard to communication between bench scientists and the world-leading drug discovery scientists on iThemba's Scientific Advisory Board. Indeed, because of the small size of the company, the scientists were able to be intimately linked with those advisers. For example, I asked one of the most senior bench scientists about their first impressions of iThemba:

"When I first heard about it, I was really taken aback by the résumés of the founders of the company. I think that was one of the most attractive features of the company. You had Dennis Liotta, you had Tony Barrett, you had Steve Ley. And I had the opportunity of meeting with these guys and thinking that these guys must be the fathers of chemistry. That was one of the most attractive features about the company." Of course, there are many world-leading scientists at Big Pharma as well, but bench scientists would be unlikely to know them. This particular scientist had worked at Pfizer in the United Kingdom, after being trained in South Africa, and reflected on the contrast between the Big Pharma experience and that at this small South African company: "I remember the CEO of Pfizer used to fly in from New York, and it was all such a big day every time he was going to meet the employees of the company. You'd be lucky to sit ten rows away from him. It's a huge difference. Getting to meet Dennis Liotta, chatting with him, or with Tony Barrett. It's a huge difference." In an international science system in which "peripheries [are] defined not geographically but in terms of scientific authority and social power," iThemba was not simply periphery.[23]

Connections between iThemba scientists and individual scientists at the center of global drug discovery created proximity in particular ways, and so did connections to processes—in this case, mixing, purifying, and analyzing substances in globally standard ways. Annemarie Mol and John Law have argued that the "network space" of laboratories renders proximity differently from a separation between a "here" (in their case, the Netherlands) and a "there" (Africa):

> In a network space, then, proximity isn't metric. And "here" and "there" are not objects or attributes that lie inside or outside a set of boundaries. Proximity has, instead, to do with the identity of the semiotic pattern. It is a question of the network elements and the way they hang together. Places with *a similar set of elements and similar relations between them* are close to one another, and those with different elements or relations are far apart. If it is actually used on arrival, then the haemoglobin meter that travels from a factory in Germany or Korea to the Netherlands and Africa turns the laboratories in both regions into a similar place. They are both "labs." They are close for they are both part of the haemoglobin-measurement network.[24]

These networks and practices mean that iThemba was in important senses not so remote from the centers of global science. Similar relations to similar machines and processes made iThemba "close" to the drug discovery labs of the Global North.

On the other hand, South Africa's geographic isolation from concentrations of the pharmaceutical industry posed real material constraints. The same scientist quoted above continued:

> I would imagine it's a lot more expensive to do research in South Africa [than in India or China], which obviously works to our disadvantage. The other disadvantage we have is the supply of reagents. Because we have reagents from Europe, it takes up to ten days before we can get many things. So that's a major disadvantage. I think what has started is to plan in advance. So if I am going to need something at the end of the month, I am going to have to buy it a couple of weeks before.

It can be easy to forget in an online, on-demand consumer environment that people in Europe and North America often take for granted, but moving things in space takes time. This is a way in which pharmaceutical research, like dynamite before it, remains tied to place. Its connection to place makes it unlike the knowledge economy of making apps. For a synthetic-chemistry-based firm like iThemba, the place of the local here points, not to any kind of authenticity, but rather to a practical accessibility. The local is of pragmatic concern.

The delays in delivery of reagents slow down South African research capacity relative to other developing countries with more robust pharmaceutical sectors, such as India and China. During my fieldwork at iThemba, this sometimes came to the fore in lab meetings in which the scientists would weigh their options regarding which of various reagents to use to synthesize desired molecules. One of the factors that often came up was whether particular components were explosive, which would add considerably to the shipping costs, which are more significant considerations in South Africa than they are in areas with more robust chemical supply infrastructures. Even if a particular material were expensive or time-consuming to make, scientists might choose it over an inexpensive and otherwise-expedient explosive alternative. This underscored that although finished pharmaceuticals are generally inert in transit, the components with which they are made can be highly volatile.

The challenges involved in securing supplies of reagents highlight the materiality of pharmaceuticals from the perspective of their developers and also the significance of material distribution in space. The advance planning needed for managing reagents is also pronounced with regard to machinery. On my first visits, iThemba did not yet have its own NMR machine, and so scientists had to take materials to a nearby lab of the Council for Scientific and Industrial Research for analysis. The acquisition of their own NMR machine marked an achievement of independence.

The availability of servicing for the machines improved over the time period of my research, even though it remained a challenge. Microbiologist Iruka Okeke points out that biomedical laboratory equipment is made by very few companies and is designed to be used close to those companies so that they can provide technical support, and "African laboratories invariably pay more for equipment but receive less service over a dramatically shortened lifespan."[25] Equipment servicing plans are prohibitively expensive or not available in regions with less laboratory density. Okeke is writing about diagnostic laboratories in Nigeria, but the observation holds for the much higher-tech laboratories necessary for drug discovery, which are reliant on equipment such as NMR imaging. There are only a handful of such machines in South Africa, and those are the majority of the ones present on the continent. The distance between these machines and their manufacturers constrains the ultimate efficacy of the machines.

That said, the distance between Europe and South Africa should not be taken too literally, even in a geographic sense. In terms of transit time, South Africa is closer to Europe than are many of the African countries between them, because of the well-developed air transportation infrastructure. There are regular flights between many rich countries and South Africa, and good roads from South African airports to the country's centers of industry. This infrastructure provided iThemba scientists with access to reagents, machines, and technicians.

This transportation infrastructure follows from South Africa's colonial history and means that the betwixt-and-between quality of iThemba as center/periphery was true at both the micro- and macroscales: virtual connection to scientists was possible, and relatively expedited transportation of materials was possible. Just as Europe itself has always had "major centers, minor centers, and peripheries" of science,[26] Africa is not merely undifferentiated periphery. South Africa, and Cape Town and Johannesburg in particular, play an important role in the geographies of those with economic and scientific power.[27]

Notably, although the scientist quoted above emphasized the importance of materials for research, they did not think that the making of drugs mattered as much as becoming the inventors and owners of IP:

> I don't think the source of the drug, the actual drugs themselves, is of any significance. I think what is important is where is a drug discovered. If a drug is discovered in South Africa and manufactured in India, South Africa still

has intellectual property rights. So you still have streams of money coming into the economy, which would help people. Obviously, money coming into the economy means the government can invest that money elsewhere. Obviously, it puts the country on the research map, if South Africa is able to discover drugs.

So here the scientist returns from the material register of the map back to a more symbolic one, yet one with real political and economic import. These comments illuminate how the scientific knowledge component of pharmaceuticals—not just their raw materials, licensing, manufacture, or distribution—matters. Even as scientists cannot ignore the ways in which highly material processes constitute a molecule that becomes the locus of pharmaceutical knowledge, once that molecule is isolated and described in the literature, it becomes invested with very open-ended possibilities.[28] The patent becomes a locus of hope, and the pride that it promises is future oriented. It ties the scientists to the molecules in a way that is profoundly aspirational.

The patriotic tenor of the comments was particularly striking because this scientist was from a different country, neighboring South Africa to the north:

Q: But didn't you say you're not South African?

A: [*laughing*] No, I'm not.

Q: Does that matter?

A: I don't look at myself as, at least when it comes to science, I look at myself more as a global citizen. As long as I am doing the research that is going to affect people's lives, it doesn't matter where I do it. But preferably in South Africa.

Q: Why South Africa as opposed to [your country]?

A: I don't think we have the capacity to do any research [in my country] right now. Maybe with time. The advantage obviously for me is that I'm much closer to home. So if I have the means, I can go [home] on a weekend and come back and be in the office on Monday. Whereas if I am in Europe, it takes many hours flying to come home.

The comments here foreground literal geographic proximity across the lines of nation-states. In this scientist's perspective, South Africa's level of development is such that building this research capacity is plausible. Since not everyone wants to live in Europe, North America, or Asia, R&D efforts in South Africa effectively expanded the talent pool of innovation.

A map does more than represent space. It also conveys how space can be navigated. iThemba's building, with its team of scientists moving between an ordinary computer room and ordinary lab spaces, might be seen as close to or far from global pharmaceutical knowledge making. iThemba was an instantiation of the hope that the global map of pharmaceutical innovation could be redrawn, such that these scientists' small lab in the outskirts of Johannesburg might have become a node in global pharmaceutical knowledge making.

Aspirations for a Knowledge Economy

The extraction of raw materials is one of the fundamental dimensions of colonialism.[29] Might a shift to a knowledge economy be part of moving more fully into a postcolonial era? What hope can synthetic-chemistry-based drug discovery provide toward realizing the transition to a knowledge economy?

Anthropologists Jean Comaroff and John Comaroff make an analogy to resource extraction when they discuss the problem of the epistemic scaffolding of Euromodernity, in which the West is the source of transcendent knowledge and the rest of the world provides only reservoirs of raw fact, for the scholarly field of anthropology and for knowledge making as a whole, "just as it has capitalized on non-Western 'raw materials'—materials at once human and physical, moral and medical, mineral and man-made, cultural and agricultural—by ostensibly adding value and refinement to them."[30] Insofar as iThemba participated in contract chemistry for pharmaceutical research elsewhere, its work was continuous with this model, albeit to a lesser degree than if it had been bioprospecting. In contract chemistry arrangements, companies or global research initiatives based in richer countries would contract iThemba to synthesize particular molecules, which the contracting companies and initiatives would treat as raw materials for further refinement and research.

For the scientists involved with iThemba, the knowledge-based goal of discovering new pharmaceuticals was often intertwined with the production-based goal of manufacturing them. Yet with both its fundamental drug discovery research and its exploration of world-leading green manufacture (the subject of the next chapter), iThemba sought a break with the colonial model, such that the value and refinement would be added at home. iThemba's goal of shifting South Africa's position from a provider of raw materials to a refiner of knowledge was incomplete, but it strove to be less one-sided.

Both wealth and prestige are at stake in being on the refinement side of the value chain. As one iThemba bench scientist put it in an interview, "if this works out, then we can actually show the international community that this is something we can do. We don't have to rely upon scientists abroad and then just get the end product, being the tablets or whatever."

iThemba represents one iteration of an aspiration to transform a developing country—in this case, South Africa—from an "extraction economy" (in which extracting minerals from the earth is the predominant source of a country's economic activity) to a "knowledge economy" (in which information becomes primary). Sociologists have defined the "knowledge economy" as "production and services based on knowledge-intensive activities that contribute to an accelerated pace of technical and scientific advance, as well as rapid obsolescence," and have suggested that "the key component of a knowledge economy is a greater reliance on intellectual capabilities than on physical inputs or natural resources."[31] The archetypal examples are science-based industries that became prominent in highly developed countries in the early 1960s—including pharmaceuticals and computing.[32]

Yet as a route to the knowledge economy, synthetic chemistry has its drawbacks. In contrast with information technology, it is far more tied to the materiality that it seeks to break free from. Moreover, R&D on an app for a mobile phone might lead to a return on investment within months, whereas drug discovery is guaranteed to take years or decades, if it succeeds at all. This is one of the key contrasts between pharmaceutical research and information technology that mattered to government funders of iThemba, who ultimately decided that their resources would have a surer return on investment in other sectors. Without high-profile already-established biotech successes, wealthy South African angel investors have followed suit.[33]

Fundamental contrasts between information technology and pharmaceutical research are illuminating. In his 1995 book *Being Digital*, MIT Media Lab founder Nicholas Negroponte famously argued that the digital era is characterized by a shift from atoms to bits: the movement of physical things matters less and less, and the movement of digitized information matters more and more.[34] Although this trajectory captures many media trends well—such as the shift from compact discs to streaming audio and the decline of newspapers—in many spheres, the importance of atoms has endured. Pharmaceuticals challenge a progressive logic of the increasing immateriality of information, through their highly material character from production to distribution to consumption.

Negroponte does acknowledge that not every industry will shift from atoms to bits:

> If you make cashmere sweaters or Chinese food, it will be a long time before we can convert them to bits. "Beam me up, Scotty" is a wonderful dream, but not likely to come true for several centuries. Until then you will have to rely on FedEx, bicycles, and sneakers to get your atoms from one place to another. This is not to say that digital technologies will be of no help in design, manufacturing, marketing, and management of atom-based businesses. I am only saying that the core business won't change and your product won't have bits standing in for atoms.[35]

Cashmere sweaters and Chinese food are remarkably low-tech examples. That passage is followed by an explanation of his target topic: "the information and entertainment industries." But newspapers and CDs are not the only things that contain information. Pharmaceuticals do too.

For Negroponte, this shift from atoms to bits is a way to move beyond the constraints of the industrial age: "When information is embodied in atoms, there is a need for all sorts of industrial-age means and huge corporations for delivery. But suddenly, when the focus shifts to bits, the traditional big guys are no longer needed."[36] Negroponte overstates the immateriality of information technology—infrastructure and labor matter in that sphere as everywhere.[37] But in the case of drugs, it is vividly clear that they will always remain simultaneously material (atoms) and informational (bits). Thus, a "digital divide" framework is inadequate for understanding the potential for synthetic chemistry in a context like South Africa. As Louise M. Bezuidenhout and her colleagues argue, "an emphasis on access fails to capture the social and material conditions under which data can be made useable, and the multiplicity of conversion factors required for researchers to engage with data."[38] What is true for chemical knowledge making is even more starkly the case for its products: drugs are effective only once they are physically materialized and bodily consumed. Moreover, products of chemistry remain profoundly material, in contrast to alternative knowledge economy paths.

The "wonderful dream" of being able to "beam up" hints at an element of the allure of the knowledge economy: the promise of frictionless movement not only for objects we desire but also for ourselves. I'll return to questions of constraints on bodily movement across borders below, but for now I want to highlight that they also impede synthetic chemistry as a path to joining a knowledge economy. It is often useful for synthetic chemists to go to particular places to learn new skills—iThemba sent promising scientists for training stints at Cambridge University in the United Kingdom and to a GlaxoSmithKline infectious-disease initiative in Singapore—and the constraints of individual scientists' passports played a role in who was selected to go. One knowledge economy dream is that knowledge makers' passports would not matter.

As I myself was traveling to and from South Africa for research, first every eighteen months and then every six months, I was constantly reminded of this distance. There is a direct flight from Atlanta to Johannesburg, which at 8,433 miles is the longest flight that the major US-based global carrier Delta Air Lines flies. The travel time is shorter from Atlanta to Johannesburg than the other way around: the flight time is usually about fifteen and a half hours going to Johannesburg and is just shy of seventeen hours returning. The difference in the flight time is mostly due to wind currents, which are very material facilitators and impediments to air travel that passengers can easily forget about unless made uncomfortably aware through turbulence.

There was something evocative about the fact that it took less time to get from Atlanta to Johannesburg than to get from Johannesburg to Atlanta. In a traditional cartographic sense, the cities are exactly the same distance apart from each other, but travel is not equally easy—and that extra hour or so in the air is the least of it. For people from the United States, all that is needed to travel to Johannesburg is about two thousand dollars for the airfare and a passport with two empty pages to accommodate the South African visa, which is issued for free on arrival. These barriers are certainly not trivial: many people in the United States do not have access to those financial resources, most Americans do not even have a valid passport, and every time I boarded the flight there were one or two people turned away for want of empty pages.[39] However, the reverse passage is far more difficult. South Africans require a visa to enter the United States, and it is challenging to qualify for. They must demonstrate "strong social, economic and/or family ties to South Africa as well as their purpose and duration of travel" and have an in-person interview at a US embassy or consulate well in advance.[40] For scientists from other African countries working in South Africa, obtaining visas could be more challenging still. Visa restrictions shaped the ability of individual iThemba scientists to go to the United States or Europe for training and for business travel. Even on this daily nonstop flight, the global flows of people are striated, not smooth.

Yet the global flows of people on transcontinental flights are just a small part of what makes the world of drug discovery far from flat.[41] The dominant archetype of the twenty-first-century knowledge economy is small groups of nimble, entrepreneurial individuals combining their education and talents to create digital start-ups. A would-be information technology start-up could conceivably be initiated by a few people operating on their own with out-of-the-box, commercially available computers and an internet connection, but drug discovery has high barriers to entry. A would-be pharmaceutical start-up needs people with highly specific training and access to wide-ranging supplies that are available through complicated markets. While an app can be released directly to consumers with minimal mediation—a couple of major phone platforms' app stores being the major gatekeepers—the path of new pharmaceuticals to market is dauntingly complex, even for major firms. Whereas hopes for a "knowledge economy" have become associated with fast-paced innovation, drug discovery is famously slow.

The necessarily material aspects of pharmaceuticals become major drawbacks to a pharmaceutical route to a knowledge economy and, with it, its decolonizing aspirations. The chemists cannot stay in the computer room, and the striated world of stuff is full of barriers and challenges. And yet pharmaceuticals' ultimate thinginess is also part of their allure: to be able to make and hold in one's hand the solution to one's own problems.[42]

Patents and Pills

IP is often understood in abstract terms, as owning intangible ideas *rather than* owning tangible stuff. This is part of why IP in pharmaceuticals can seem so intuitively unjust: why should ownership of ideas among the few be a barrier to access to urgently needed, actually existing drugs among the many? Yet it is worth bearing in mind that, on many theoretical and practical levels, pharmaceutical ideas and chemical stuff are inextricable.

Curiously, notions of IP have themselves been shaped by chemical materiality. From the inception of industrial-age IP, chemical ideas have been drawn upon to understand inventiveness. Mechanistic understandings of patentable machines went hand in hand with chemical understandings of patentable processes and products, as exemplified by vulcanization, which is the chemical process of using sulfur and heat to convert natural rubber into a more durable material. Alain Pottage describes the role of analogies to chemical reactions in patent law in this way: "the image of chemical reactions as synergistic processes emphasized the creative act of bringing components together: the invention lay in the (act of) relation, and the identities of the components were relative and emergent rather than inherent and predetermined."[43] If chemical metaphors of emergent combinations can inform the patentability of a broad range of technologies, the chemical space itself never loses its physicality.

From the perspective of a pharmaceutical company, both patents and pills are highly desirable things. The difficulties in making both novel IP (i.e., patents) and distributable pharmaceuticals (pills) include differently materialized aspects. As a drug discovery company, iThemba's mission was more focused on making patents than pills. That said, many of the various business models that iThemba Pharmaceuticals pursued explored possibilities of pharmaceutical manufacture as a means of securing cash flow and of appeasing the company's South African government funders' desire for useful things in the short term. In these models, the manufactured pharmaceuticals would be generic ones that would serve local and regional markets, especially essential off-patent antiretrovirals such as tenofovir, and the cash flow would fund drug discovery research that would ultimately result in novel IP and a patentable drug.

The images in figure 7 are both depictions of Emtriva (emtricitabine), an antiretroviral drug discovered at and named for Emory University by iThemba cofounder and scientific adviser Dennis Liotta and his collaborators. The image on the left is of the molecular structure, and the image on the right is of a brand-name pill. In what senses are these images of the same thing, and in what senses are they images of different things?

One gap between what the images represent is certainly branding. But another is their materiality. Very specific lab-scale material processes precede the molecular structure depicted in the image on the left, and very specific industrial-scale material processes precede the pill depicted in the image on the right. Intersecting but distinct bureaucratic processes facil-

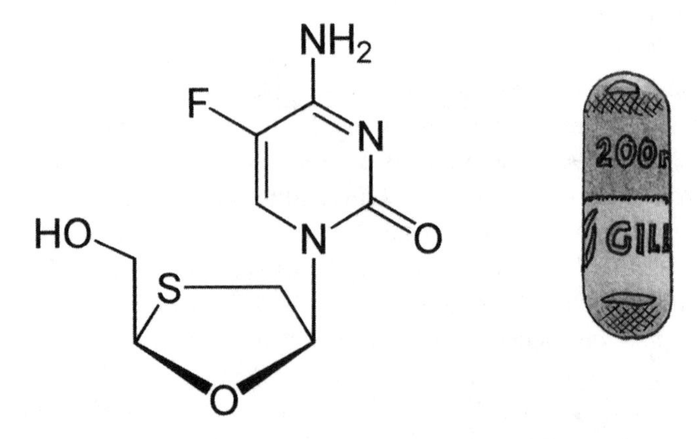

7. Two forms of Emtriva: molecular structure (public domain) and pill (drawn by Hannah Dar).

itate each materialization, of patents and of pills: regulatory bodies and inspection practices,[44] international trade agreements and complex supply chains. Both of these sets of materializations have rich histories,[45] and both are important for understanding the promise and challenge of creating a pharmaceutical industry in South Africa with the capacity to both invent novel drugs and to manufacture drugs, whether patent protected or generic.

Emtriva, the pill, is actually rarely sold as itself—it is more often delivered in a multidrug pill, such as the once-a-day Atripla that is the preferred treatment for HIV by the US Department of Health and Human Services. From the perspective of a patient, the handful of pills that make up Atripla both are and are not the same thing as a single, once-a-day pill.[46] We can acknowledge the informational truth of multiple pills being *the same* as a combination pill, but experientially, a lessened "pill burden" matters. And this highlights the fact that pharmaceuticals are necessarily material not only from the perspective of producers but also from the perspectives of consumers. The information that pharmaceuticals instantiate is only truly transmitted after the highly processed chemicals pass through the consumer's lips, throat, digestion, and physiology—pill taking is a key site of traffic between information and flesh.[47]

In global health discourse, much more attention has been paid to patent barriers to pharmaceutical production than to other material barriers. There is a widespread tendency in access-to-medicines movements to operate on the assumption that if the IP of a useful drug is unrestricted by patents, the drug can and will be made cheaply. This affordable manufacture does sometimes happen, as we saw in a previous chapter: the fight for access to HIV drugs in South Africa and elsewhere in the late 1990s in which Indian generic manufacturers became vital suppliers of urgently needed antiretrovirals.

However, there are many barriers to making any drug, whether or not it is patent protected. This has been underscored in scandals around entrepreneurial pharmaceutical profiteering in the United States. For example, in 2015 Turing Pharmaceuticals, led by "pharma bro" Martin Shkreli, raised the price of a decades-old toxoplasmosis drug by 5,000 percent overnight, and hospitals and payers had no choice but to pay because there were no other companies producing the drug.[48] Indeed, there were no companies with the capacity to enter the market quickly. This shows that to conflate access to a drug patent with the capacity to actually make a drug is an error.

Failure to consider the distinctive material barriers to pharmaceutical production leads to misguided comparisons between IP in information technology and in pharmaceuticals. Consider this absurd statement printed in the *New York Times* in 1998, discussing a proposed law that would allow South Africa to import generic antiretroviral drugs and eliding the differences between copying drugs and copying software:

> "If the Health Minister thought it was in the interest of public health that those $10,000 AIDS cocktails be cheaper, she could just rip off the patents and set up a factory in Cape Town to make them," said a Western diplomat who is fighting the law. "And if the Minister of Health says this is O.K., then the Minister of Education will be able to say, 'Well, affordable computers are in the interest of public education, but Windows is just too darn expensive, so we're going to buy knockoff copies.'"[49]

In these articulations, the process of materializing pharmaceutical IP is treated as trivial. But copying drugs and copying software are radically different, in ways that matter.

South Africa has relatively little pharmaceutical manufacturing, much less R&D. In order to explore the possibilities of building a robust pharmaceutical sector in South Africa, it's worthwhile to note that there are many barriers to entry, beyond IP, including access to reagents, equipment, electricity, and water; trained scientists and other workers; processes of quality control; and the ability to navigate regulatory environments.[50] Traditional pharmaceutical manufacturing also requires large facilities. Relative to other parts of Africa, South Africa does have significant elements of the infrastructure necessary for pharmaceutical manufacture and even process innovation and drug discovery. It has well-regarded universities that produce a good number of scientists, an IP environment that protects producers, and effective power and transportation infrastructures (for the country's industries if not for the country's population). The infrastructure of academic and industrial science in South Africa is that of a developed country, albeit not a leading one and not very diversified, since it is dominated by a single sector, mining. South Africa also has a significant middle class and relative political and social stability. The social and economic infrastructures create conditions of possibility for the production of both pharmaceutical patents and active pharmaceutical ingredients and pills.

That spatial distribution of pharmaceutical infrastructure is itself an instantiation of legacies of regional and global political and economic orders. These spaces are always-already part of a network, most immediately because synthetic chemistry makes ample use of commercially procured "intermediates" rather than starting from first principles. No one does synthetic chemistry alone.

Both patents and pills are sites of hope: controlling patents offers particular kinds of power, and the ability to make pills and their components does too. Pharmaceutical science has more prestige than pill manufacturing does, but both can be understood as routes to development distinct from the core of the South African economy in mineral extraction. As I will explore in the next chapter, the im/material project of drug discovery at iThemba would become intertwined with the im/material project of innovation in pharmaceutical manufacture.

SIX

Hope in Flow

As we saw in the previous chapter, the rhythm of the days of the bench scientists at iThemba Pharmaceuticals was characterized by distinctive sets of movement: scientists moved back and forth between the computer room and the labs, and within the labs, they transformed materials back and forth between powders and liquids. However, toward the end of the period of my research, scientists there were also investigating possibilities of an alternative rhythm for the lab work, one driven by an effort to develop the capacity for iThemba to engage in an ascendant model of chemistry production known as continuous-flow chemistry, in which materials are kept liquid throughout the reactions and isolated into powder only at the end. This innovative way of materially engaging with the chemicals would hold hope for positioning the company and South Africa in new ways in the broader terrain of the global pharmaceutical industry.

This book has so far focused on how the scientists at iThemba hoped to join and participate in the world of drug discovery—their company was founded with the mission of finding new drugs for TB, HIV, and malaria. Here, I explore how the company's focus widened to include hope in process innovation: manufacturing existing drugs in novel ways. In this endeavor, iThemba sought not only to achieve parity with global standards in pharmaceutical manufacture but to leap ahead. In what I will describe as a postmodern post-apartheid parallel with scientific innovations during the apartheid era, the conditions of possibility for world-leading innovation were inextricable from conditions of constraint and lack. There are echoes of South Africa's apartheid-era isolation from petroleum production in the country's current marginalization from global pharmaceutical production. Exclusion created and now creates spaces of possibility for doing things differently. However, the neoliberal post-apartheid context in contemporary South Africa has made it very difficult to realize the promising potential in this pharmaceutical space.

In this chapter, I meditate upon the hope in flow chemistry and am especially attentive to temporality and infrastructure. Science and technology studies have generally been more successful at accounting for space than at accounting for temporality, and both are at stake here.[1] Flow chemistry itself provides an opportunity to reconceptualize time in pharmaceutical production, since its material constraints and affordances require changing the order of chemical reactions. Situating flow chemistry in South Africa adds macroelements of time as well. Among the scientists involved, there was a sense that South Africa is "too late" to the pharmaceutical production game to compete with India and China in standard "batch" pharmaceutical production techniques, but is not "too late" to lead in flow. South Africa's lack of robust infrastructure for the pharmaceutical industry's current highly toxic manufacturing practices becomes, in this aspiration, a condition of possibility for leading in the green manufacturing practices of the future.

Copying India's Model Was Not an Option

iThemba Pharmaceuticals was founded as a drug discovery company, with a mission of finding new drugs for TB, HIV, and malaria. It was not founded as a vertically integrated pharmaceutical company with both research and manufacturing capacity. Since drug discovery and development are famously slow, it was a mission that operated on the assumption of a long time delay between research input and product output. Yet over the course of my five years of research at iThemba, between 2010 and 2015, South African government officials from agencies that had provided iThemba's start-up funding wanted to begin to see a return on their investment, and they urged the company to start earning its keep by making useful, sellable products in the short term. This increasingly became an imperative, a condition imposed if the company were to continue to receive government funding.

In light of the high burden of HIV in South Africa, an obvious example of a useful, sellable product for iThemba to make would be generic antiretrovirals. However, it would not be easy to make such an enterprise competitive, since low-cost Indian manufacturers had a strong hold on that market.[2] Globally, the procurement of generic antiretrovirals involves a complex network of donors and governments.[3] During the period of my research, the South African government itself procured a major portion of the antiretrovirals consumed by the South African people, and its policies placed a high priority on obtaining these essential drugs at low prices.[4] As health technology innovation analysts Palesa Sekhejane and Charlotte Pelletan have argued, the South African government "has no clear trans-sectoral strategy to enhance indigenous innovation," and so even though the Department of Trade and Industry has guidelines that are supposed to privilege local manufacturing over low-cost imports, the Department of Health, which does the actual procurement of antiretrovirals, ignores that policy in favor of purchasing drugs at the lowest possible cost.[5] Many aid programs that provide antiretrovirals also have guidelines to encourage purchase from the lowest bidder. Any new entrant would have a very difficult time matching the low prices of well-established generic manufacturers.[6] How might South African antiretroviral manufacture find its place in local and global marketplaces?

In the several years that I spent speaking about the research for this project, many people have asked me why South African companies don't just copy India's model: develop a robust generics industry through reverse engineering branded drugs from the North. This could serve the immediate needs of the people, and profits generated could fund research toward innovative drugs as well.[7] However, South Africa's path to a more robust pharmaceutical sector could not simply follow the route taken in India.

Part of the reason that South African companies needed to pursue a different path from Indian companies is due to global intellectual property regimes. The large Indian generic-pharmaceutical industry is the product of a particular history and was enabled by the specific arrangement of international trade agreements in which India was granted more flexibility than South Africa was.[8] Moreover, that early advantage has now been consolidated, such that Indian companies are now positioned as global leaders of generic manufacturing in low-cost environments.[9] Now that the generic-pharmaceutical industry is so well developed in India and a handful of other countries, this model is even harder for new entrants in new places to replicate. Well-established companies can withstand a very small profit margin, or sell particular products at a loss, in order to undercut potential new competitors and protect their market share. It is "too late" to beat the Indian generic manufacturers at their own game.

At the same time, from the perspective of would-be participants in the global pharmaceutical industry, replicating the Indian pharmaceutical industry is not necessarily the ideal. India's pharmaceutical sector is perceived of as second-rate, relative to companies in the historic innovator countries such as the United States, Switzerland, and Japan, and they provide African consumers with drugs held to lower standards of quality than the drugs they supply to American consumers.[10] For the bench scientists at iThemba, the association between India and inferior quality extends to the reagents needed to do synthetic-chemistry research. At a lab meeting that I attended in 2012, the bench scientists were talking through a plan for a method of synthesizing a target molecule with a visiting Scientific Advisory Board member. When they got to a particular step, the visiting scientist probed: how will you make that reagent for that reaction? The scientist who had been speaking answered, "I will order it from India." After a beat, the scientist continued, "And when it turns out to be sugar, I will order it again." The whole room laughed. The relationship with India is a vital one, for synthetic chemists as well as for pharmaceutical consumers, but there is scant trust. In this sense, it's just as well that it is too late to occupy Indian companies' niche—South African companies hope to be able to snag a better one.

The reliance on India for supplies of reagents for research is continuous with South African manufacturers' reliance on India for active pharmaceutical ingredients (APIs) for drugs that are formulated and manufactured locally. Imported inputs are inherently risky for local manufacture, for many reasons. First, as suggested above, foreign companies can use their control over the price of key inputs to harm potential local competitors, either by raising the price of key inputs to make the production process price prohibitive or by lowering the price of the final products to make production unviable for potential competitors. Second, when supplies are short, the producers generally supply their more important markets first (i.e., the United States and Europe), and less lucrative markets bear the brunt.[11]

Since India shifted to compliance with the Trade-Related Intellectual Property Rights Agreement in 2005, Indian companies have increasingly developed the capacity to do R&D on innovative drugs. However, that R&D has been largely directed toward the global market rather than toward diseases of the poor in the Global South (such as TB and malaria).[12] As anthropologist Kaushik Sunder Rajan argues, the "harmonization" of global IP laws to bring India into the fold "must be understood in terms of expansion of multinational corporate hegemony."[13] If the priorities of Indian pharmaceutical companies are shifting away from those companies' historic role as "the pharmacy of the developing world," and yet they can prevent companies elsewhere from taking up that role, what other routes toward pharmaceutical manufacture by and for the Global South might be available?

If iThemba were to start making generic antiretrovirals, it would have to find some way to distinguish itself in the global market. iThemba's leadership saw a potential niche for the company in novel-process chemistry, specifically a "green chemistry" process known as continuous-flow chemistry. Manufacturing already-existing antiretrovirals and other drugs in this innovative way was imagined as a worthy enterprise in its own right and also as a way to gain the cash flow necessary to support the company's founding mission of drug discovery for TB, HIV, and malaria. It would involve implementing the expertise of Steve Ley, a member of iThemba's Scientific Advisory Board who is based in Cambridge, United Kingdom. Ley is a global leader in flow chemistry, a hot trend in fine-chemical production that promises to radically lower the quantity of solvents, and thus the environmental impact, of pharmaceutical production. This process has other potential advantages as well, including greater efficiency at small scale and a smaller footprint (less land intensive). The idea was to build "green" pharmaceutical production infrastructure and capacity in South Africa.

Rather than copying India's model, the idea with flow manufacture was to skip India's model. We might read optimism in flow chemistry as an analogue to the optimism about mobile phones' capacity to allow Africans to "leapfrog" telecommunication technologies, skip steps, move ahead developmentally without building cumbersome landline-based telecommunications infrastructure first.[14] As science and technology studies scholar Toluwalogo Odumosu has argued, this articulation elides a great deal of complexity in how Africans make telephony their own.[15] Yet the aspiration to leapfrog has a long history in science and technology policy in the periphery, where elites have been drawn to prospects of development that do not require thorough social and institutional change,[16] and the promise of escape through avoiding waste might have been particularly appealing in a historical moment in which South Africa was reeling from the so-called "poo wars," in which theatrical protests have decried lack of access to basic sanitation in many parts of the country.[17] Following a linear path from this level of abjection can seem unacceptably slow to political leaders and the mass of population alike.

"Leapfrogging" is a business idea as well as a statist and popular one. The emergence of a world-leading pharmaceutical industry in India in the 1990s has been described as leapfrogging, and even the emergence of the pharmaceutical industry in the United States in the 1930s has been described this way.[18] Business history is replete with examples in which an important but underdeveloped sector in one country has risen to global prominence by overtaking the sector's current leaders through well-chosen technology shifts that open up the next era.

Dreams of leapfrogging have intensified in a postmodern era,[19] and with mobile phones before (or simply without) landlines, this elusive promise of leapfrogging in Africa has gained credibility. Leapfrogging, skipping ahead, is like finding a "wrinkle in time": picture an ant walking along the hem of a skirt; the distance it must travel is radically shortened with a well-placed fold.[20] Might flow chemistry provide that fold in a trajectory toward a pharmaceutical future? iThemba sought funding from the South African government to import the necessary flow chemistry technology from a European manufacturing partner, but that funding ultimately did not come through. Thus, the hope for flow chemistry would not be realized at iThemba.[21] Yet the idea of developing flow chemistry pharmaceutical manufacture in South Africa remains a useful one for thinking through pharmaceutical geographies.

The Process of Flow

In an interview with me at his lab in Cambridge, United Kingdom, in 2014, iThemba Scientific Advisory Board member Steve Ley, arguably the world's foremost academic expert on flow chemistry, explained its advantages. Describing the history of chemistry, he said, "Traditionally, we worked in a flask. We have done it that way for hundreds of years. We have lots of recipes, it works, and it is the basis of modern society. But it creates a lot of waste." As a model for doing things more efficiently, he pointed to the cell. "It's been doing it for four billion years: making its own catalysts, recycling reagents, doing continuous molecule processing." He wondered, "can we sit somewhere in the middle?" He was making a case, not for biotech processes using cells to grow desired molecules, which would harness biological reproductivity for biotech productivity,[22] but for a more efficient chemistry approach inspired by the cell's efficiency, with continuous molecule processing.

When a couple of Ley's postdocs showed me around the various labs, they presented the techniques being used with notable pride and spoke fondly of the iThemba scientist who had come to their lab on secondment to learn about their methods. There were many rooms of lab benches under standard hoods but with a distinct look: far fewer Erlenmeyer flasks and, instead, many long and very narrow tubes containing the continuous reactions, monitored by computers.

This visible difference is a theme that Ley and his students and collaborators have highlighted in their publications: "A significant contributing factor to this wastage is the relatively conservative nature of the chemical community. Unlike many other sciences synthetic methodology and laboratory practices have remained relatively unchanged for a long period of time. A chemist of 50 years past would still recognise much of the general apparatus, tools and equipment in use within a chemical laboratory today. Our reliance on batch processes has shackled us and engrained a dependence on round bottomed flasks, separating funnels and other traditional glass manifolds."[23] This language of "shackles" is hyperbolic but evocative. The invocation of freedom is resonant with the promise of leapfrogging mobile phones—perhaps it is time to leave the nineteenth-century tools of Erlenmeyer flasks and wired telephones behind and move into the twenty-first-century future untethered.

The basic differences between "batch" and "continuous-flow" processes are straightforward to understand but wide-ranging in their implications. Consider a toaster as an analogue. A "batch" toaster is the kind that most people have at home: it starts a new toasting process each time it is turned on and makes toast two or four slices at a time from beginning to end. A "flow" toaster is more like the one at a restaurant or hotel breakfast bar, a "conveyer toaster," with the machine constantly running and toast moving through. Conveyer toasters are preferred in commercial settings because they have higher throughput. iThemba's government interlocutors sometimes conflated product and process—referring to the flow chemistry technology that the scientists wanted to procure by the acronym of the API that they were proposing to make. This might have followed from an intuitive logic in which toast and toaster are closely linked, similar to calling a toaster a "toast machine": not something that a native speaker would say but comprehensible. Yet the flow equipment could in theory be used to make a wide array of molecules, and the funders' error in appellation was frustrating for the scientists, a stark demonstration of how hard it was to get the government liaisons to understand what they were trying to do.

Of course, the chemical reactions of pharmaceutical manufacture are more complicated than that involved in toasting bread. The only chemical reaction in toast is the dextrinization of starch, whereas pharmaceutical manufacture involves mixing and combining various substances that change state and temperature along the way. One reason that the process efficiency of pharmaceuticals has historically been low is that "most pharmaceuticals are very complex organic molecules that have to be constructed using multiple synthetic steps, often involving the isolation and purification of intermediate products."[24] Whereas all the bread in the toaster will ultimately become product, only a tiny portion of what goes into a batch pharmaceutical manufacturing process will become product.

To make the same end product, the sequence of the reactions themselves is very different in traditional "batch" manufacture and "flow," because the ideal nature of the material along the way is opposite: crystalline versus liquid. In batch manufacture, each reaction or series of reactions is followed by a purifying step—removing the solvent and by-products and saving the desired material for the next reaction. Thus, it is ideal to have pure, solid, crystalline material produced at each step. In flow manufacture, all the reactions are set up in a continuous sequence in the machine, and the purification of the product does not happen until the end. Accordingly, it is instead ideal to have no crystals formed along the way.[25] So, in batch, ideally there are solids and easy-to-remove liquids, whereas in flow, ideally there are liquids the whole way through the process right until the end.

Some of the implications for pollution in the shift from batch to flow are easy to see. Roughly 75–80 percent of the waste produced in the traditional process of pharmaceutical manufacture is solvents,[26] and flow requires much less solvent. Traditional batch pharmaceutical manufacture produces radically more waste than product. For example, consider tons of waste per tons of product. Compared with the oil industry, the pharmaceutical industry produces 250–1,000 times as much tons of waste per tons of product.[27] One way of understanding this inefficiency is the idea of "process mass intensity," which is the total mass used in a process or process step divided by the mass of product produced. There are other aspects of "green chemistry" and other ways of benchmarking it, including reduction in life cycle energy use and carbon footprint,[28] but process mass intensity is the standard promoted by the American Chemical Society Green Chemistry Institute's Pharmaceutical Roundtable,[29] and it is directly addressed by flow processes.

Apart from decreasing waste, changing the sequence of reactions when shifting from batch to flow has other implications as well, including a significant decrease in explosiveness. Because in flow fewer molecules are touching each other, temperature is more tightly controlled, and reactive material is continually processed rather than accumulating, there is much less chance of combustion.[30] This is particularly striking in the case of iThemba because the company's site was on the edge of the grounds of a historic dynamite factory (see chapter 2). The prospect of flow highlights a way in which explosiveness remains relevant in the physical space of pharmaceutical manufacture. As end products, pharmaceuticals are ideally inert until ingested, but along the way, many APIs, solvents, and excipients are explosive—something that pharmaceutical manufacturers need to take into account in their operations.[31]

Batch processes also require a large, fixed physical footprint for a manufacturing facility, whereas flow processes can be set up in intermodal shipping containers—the familiar structures that can be moved by ship, train, or truck without unloading and reloading their cargo.[32] These form a modular infrastructure that can be easily moved and set up in parallel. Intermodal containers are icons of today's global supply chains,[33] which makes manufacturing that can be performed in them (or something like them) intuitively appealing. Since "the development of contained and controlled facilities enables product standardization within the pharmaceutical industry," the prospect of doing so in such a flexible form holds a compelling appeal.[34] The container becomes to a significant degree its own infrastructure, demanding less from its surrounding facilities and terrain.[35]

When members of iThemba's Scientific Advisory Board gave presentations on their flow chemistry business model, contrasts between large manufacturing plants and shipping-container-sized systems were enthusiastically foregrounded. Like the freedom from the "shackles" of traditional chemistry glassware that Ley invoked when describing the way that his lab worked, the prospect of manufacturing in mobile containers offered the promise of freedom from traditional factories. Invocations of freedom and mobility became grounds for raising the possibility of pharmaceutical manufacture in previously marginalized South Africa.

Pollution and Possibility in the Supply Chain

Flow chemistry is a key part of "green chemistry." The word "green" itself is intertwined with hope because of its association with spring and new beginnings, forward movement, and environmental sustainability. For most of the twentieth century, chemical industries responded to environmental concerns with dismissal, but by the 1980s, "stewardship became the name of the game"—especially in terms of public relations but also shaping science and advocacy.[36] Even as the industry broadly has shifted toward emphasizing benefits and disavowing environmental risks,[37] promises of sustainability still have an allure. This is especially true for the scientists who have been advocates of green chemistry, who do not see themselves as participating in "chemical industry greenwashing" but rather as responding to a genuine desire to address their industry's role in environmental degradation.[38] These kinds of green ideals can complement visions of a just postcolonial science that drive efforts to foster African drug discovery capacity. They can also complement a more fundamental ideal in engineering of developing ever more efficient processes, which are more elegant and can lower costs for raw materials, waste management, transportation, and disposal.[39] What would it mean for iThemba to innovate in green space, and how might that create a space of competitive advantage?

Historically, the financial incentives for Big Pharma to introduce green chemistry methods have been very weak, because the overall margin on pharmaceuticals has been huge but the cost of producing the API is a tiny part of the cost of the final drug. API manufacture has long been generally low-cost and low-margin—the margin is made later, in the final formulation. This is in contrast to fine-chemical manufacture for the electronics industry, for example, where the cost of the components is a high portion of the overall cost in a high-margin field. Whereas pharmaceutical manufacturers have generally sought to maintain or increase margins by focusing on high-price innovative drugs, fine-chemical manufacturers have had an incentive to reduce manufacturing costs.[40]

In this sense, the progression to flow chemistry extends from iThemba's sideline practice since it began: contract chemistry. iThemba had long been relying on contract chemistry for a significant portion of its cash flow (the rest being funded by the South African government's Technology Innovation Agency). This contract chemistry involved producing molecules and intermediates for companies in Europe and North America for a fee. Thus, iThemba had incentive to make the production of those molecules and intermediates efficient. The shift to flow chemistry would have allowed them to make chemicals at a larger scale (kilograms rather than grams, industrial scale rather than lab scale). The hope was then to also reap a higher margin by making APIs and controlling the value chain through formulation.

Insofar as there is public awareness of pollution due to pharmaceuticals, it is generally about the *products* rather than the *process*. This is similar to discourse around hybrid cars, in which the pollution caused in the process of manufacturing new cars is obscured by focus on pollution that occurs during their use. For example, when the WHO quantifies "waste from health care activities," it includes "pharmaceuticals: expired, unused, and contaminated drugs; vaccines and sera."[41] Even when pharmaceutical scientists write about the pollution in the industry, they tend to put much of the emphasis on discarded pills.[42] That is where pharmaceuticals are at their most tangible, but it is not necessarily where most of the pollution takes place. In the supply chain, much of the environmental impact is from solvents and excipients.[43]

Another reason that there has been little incentive for Big Pharma to change its manufacturing practices is the fact that so much API manufacture, even for North American and European companies, happens in India and China. If the regulatory environment is such that companies don't have to pay for the externalities of disposing of the toxic waste properly and can just dump it in local waterways, the cost of manufacture that actually makes it to the balance sheet is kept artificially low. But environmental pressures may play an increasing role in the pharmaceutical supply chain.[44] There is a chance—particularly in the European Union—that manufacturers will be required to account for their impact on the environment even if it is in India or China. If this happens, there will be a huge incentive to make processes more efficient. "While regulatory pressure mounts, so does economic pressure: pharmaceutical production is vastly more inefficient than other industries and often relies on hazardous materials, making GC [green chemistry] innovation potentially very lucrative."[45] This hope in a potential greener future for the pharmaceutical industry is where iThemba looked to find its competitive advantage and to put South African pharmaceutical production on the global map.

In general, process innovation toward cleaner and more ethical products is conceived of with wealthy consumers in mind, whereas products for poor consumers are produced with lower health and safety standards.[46] One reason that flow chemistry methods seemed possible is that wealthy purchasers play a significant role in one sense of the consumption of antiretrovirals—as payers—even when consumption in the literal sense of ingestion of pills would be by the poor.

Greener pharmaceuticals provide a bridge between the now significantly discredited but still resonant twentieth-century promises of "better living through chemistry"—to draw on a widely known variant of the US transnational chemistry giant DuPont's slogan adopted in 1935[47]—and more contemporary political economies of hope located in biotechnology. A moral economy of hope in biotechnology feeds financial investment in a political economy of hope directed toward that space globally,[48] and a central idea of using flow chemistry was to achieve that simultaneously affective and financial effect of investment in a better future.

Post-apartheid Postmodern

There is something evocative about the fact that this vision of starting afresh represented by hope in flow chemistry involves a play between solid and liquid. South Africa has long built plants to turn solid coal into useful liquids, including ammonia, methanol, and synthetic natural gas. Coal-based plants of this type are only "viable under special circumstances, where coal is cheap, gas is expensive or unobtainable, and where geography makes it expensive to import nitrogen fertilizer from elsewhere."[49] Because apartheid-era South Africa had abundant coal and cheap labor and lacked natural gas and petroleum, and its distance from suppliers made shipping potentially explosive fertilizers expensive, it was an ideal site for such production. Factories in Modderfontein itself, the part of Johannesburg where iThemba was situated, have been world leaders in this sphere.

Apartheid South Africa, driven by the desire for fuel independence in the context of global sanctions, took the innovative step of realizing the dream of creating liquid fuel from coal—something that the state played a key role in making possible, with a capital outlay in the state corporation Sasol far beyond what private industry could accomplish on its own.[50] Coal-to-liquid technology and Sasol as a state corporation are important in what historian Stephen Sparks characterizes as "'apartheid modern,' the particular form that industrial modernity took under apartheid."[51] Its form of nationalism celebrated the ingenuity of South African scientists and the turning of rural Afrikaners into industrial citizens, while rendering invisible the cheap black labor and corresponding artificially low energy costs on which the enterprise depended. This enterprise was integral to the effort to make South Africa self-sufficient and thus able to perpetuate its social structure in violation of international norms—without vulnerability to lack of access to petroleum that sanctions could present.

Flow technology presents a potential post-apartheid analogue, one that embodies hopes of creating what we might call a "post-apartheid postmodern." Whereas hope for modernity was a vision of moving to the next stage of development in a teleological developmental model, postmodern progress would be realized through nonlinear development and participating in disruption. Although the word "flow" has resonances with continuity, it becomes a site of hope in discontinuity.

In flow chemistry as with industrial modernity under apartheid, scientific ingenuity remains key, but all other labor becomes almost completely invisible. Indeed, in flow chemistry, labor is reduced relative to traditional batch manufacturing processes, which is part of flow's appeal. Leading flow chemistry scientist and iThemba Scientific Advisory Board member Steve Ley considers it an additional waste reduction that flow chemistry offers: "it is increasingly unacceptable to waste the human resource on trivial or repetitive scale-up tasks."[52] The "human resource" (labor) that he is referring to is exclusively of the scientifically trained, highly skilled technical type that correlates with the scientific ingenuity in the mining sector under apartheid. Labor of the skill level that miners would have is almost nonexistent in Ley's imagined future, and completely invisible.

There are some stark contrasts between the apartheid-era and post-apartheid cases, especially with regard to the environment. Whereas coal-to-liquid technology had been tremendously polluting, the appeal of flow is the appeal of the clean. The fact that environmental impact figures in the design of flow—rather than pollution being mitigated only after the fact or even tolerated and ignored—contributes to the sense that it fits into our contemporary era. In this figuration, South Africa would not be isolated and self-sufficient but integrated and cutting-edge.

The technology for this route to a post-apartheid postmodern cannot sustain a fantasy of total self-sufficiency. It would draw on outside expertise. One of the scientists from iThemba was sent on secondment to Ley's lab in Cambridge for training, and one of the postdocs in Ley's lab was sent to iThemba for three months of industry placement. The proposed model also relied on collaboration with a Swiss pharmaceutical company. Yet it would not be merely derivative of technologies elsewhere. Ley's lab's most high-profile successes in flow were with small-volume high-margin molecules such as Imatinib (the API in the infamously expensive leukemia drug Gleevec),[53] whereas iThemba's specific focus would be on antiretrovirals. Moreover, since the pharmaceutical infrastructure would be new in South Africa rather than a small supplement to an already-existing infrastructure, South Africa would be positioned well ahead of global standards.

If flow were to be realized, the state would likely again be key, as it had been in the coal-to-liquid enterprise. The state figured not only as initial investor but also as market. One of the first products that iThemba planned to produce using flow technology was a key generically available antiretroviral API. This is not necessarily the most logical first product to pursue, if thinking in a global context, because antiretrovirals are so low margin. And yet it would become feasible if the market could be guaranteed in advance. Since the South African government is a major purchaser of antiretrovirals, it might be possible to negotiate a tender even if the drugs were not initially able to be produced more cheaply than they could be abroad. This would make South Africa more self-sufficient, less reliant on the vagaries of world markets. Relying on private Indian pharmaceutical manufacturers to provide generic antiretrovirals puts South Africa in a vulnerable position. As Indian manufacturers have come into fuller compliance with the Trade-Related Intellectual Property Rights Agreement since 2006, they are increasingly able to move more fully into higher-margin markets, and there is no guarantee that providing antiretrovirals to Africa will continue to be a priority.[54]

However, the scale of innovation that flow chemistry for pharmaceutical production represents is much harder to achieve in the neoliberal context than it was in an earlier era. Under apartheid, South Africa was a *liberal state*—albeit only for a minority of its citizens. Post-apartheid South Africa is a *neoliberal state*. In a liberal model, state building and economic development are long-term projects. A liberal state can invest heavily in innovation without paying much attention to when and where the returns on investment come. In contrast, a neoliberal state has to constantly worry about return on investment. There is a particular cruelty to the fact that investment in R&D was plentiful under apartheid but is scarce under democracy, which resonates with the sense of loss in the "broken tempos" elsewhere in Africa where lab scientists have nostalgia for a (relatively) well-equipped colonial past.[55] The blatancy of the oppressive nature of apartheid-era state-supported science makes nostalgia for "glory days" of South African science untenable,[56] even as it makes the present situation even harder to bear. At a time in which the rainbow nation is at the bench ready to innovate, the rules of the game have changed to make it much harder to secure the necessary resources.

The challenges in securing the necessary capital investment notwithstanding, in light of its distinctive historical and regulatory situatedness of a strong scientific infrastructure and developed-country intellectual property regime, South Africa is well positioned to build pharmaceutical manufacturing infrastructure, but at the same time, it is not encumbered with capital investment in the industry's current manufacturing processes. Insofar as flow chemistry allows "disruptive innovation," those who are not already invested in the current practice are at an advantage. Lack of infrastructure becomes freedom from "infrastructural lock-in." Consider an analogy: the existence of an urban transportation system designed for individual cars can increase the barriers to creating a more efficient transportation system for that city. Once inefficient infrastructure is in place, it is very expensive and difficult to change to a better model. In the case of pharmaceutical manufacture and the hope in flow in South Africa, the conditions of loss in the past and absence in the present become conditions of possibility for a future that skips a step and starts ahead, afresh, anew. Yet the need to think short-term renders the realization of that hope tremendously difficult. This makes government investments in information technology innovation far more appealing than pharmaceutical investments, because an app can be sold within months of its start, whereas implementing flow manufacture of pharmaceuticals would take years. The government funders do not necessarily have that kind of time.

At stake here, then, are aspirations that are far more local than a general desire to reduce the pharmaceutical industry's global environmental footprint. As with much of what iThemba has ventured to do, the aspiration to implement continuous-flow chemistry can be read as a "sociotechnical imaginary"—a vital site for thinking through not just what technology should be but also what the nation should be.[57] The hope in flow enrolls a hope to create a "place in the world" for African science—in this case, South African science.[58] Africa has long been framed as a *recipient* of pharmaceuticals rather than a *producer* of them, even in global health discourse that seeks to expand access.[59] For example, Doctors without Borders literature often describes India as "the pharmacy of the developing world."[60] In this articulation, there is a sense that the places of the pharmaceutical industry are already fixed: innovation will occur in Europe, North America, Australia, and Japan; copies of those innovations will be made in India, China, and Brazil; the rest of the world will have no role at all as producers, only as consumers of what those few places decide to produce and share. Scientists in rich countries play the role of subjects of science, those in a few middle-income countries play the role of intermediaries, and those in the rest of the world are relegated to objects insofar as they are not excluded altogether. There is a hierarchy underlying this geographic division, in which Africa can never be its own pharmacy, much less participate in the enterprise of producing pharmaceuticals for export.

The hope in flow chemistry is, in a sense, the hope that it is not too late for the global map of pharmaceuticals to be redrawn. This hope in flow represents an optimistic engagement with time in Africa. Achille Mbembe has pointed out the tendency in writing about Africa to "assimilate all nonlinearity as chaos, forgetting that chaos is only one possible corollary of unstable dynamic systems."[61] Amid broader disillusionment around ideas that Africans are on a linear path of progress toward modernity, what alternatives to chaotic stasis or disintegration exist?

EPILOGUE

The Afterlives of Hope

On one level, this book has told a very specific story: the contexts and accounts of a small South African pharmaceutical company called iThemba—a name that is Zulu for "hope"—its emergence and ultimate failure. Throughout, I have also engaged broader themes of place and knowledge making, using this particular site to ground wide-ranging consideration of histories, infrastructures, politics, and hopes. Studying this case illuminates the limitations of global health frameworks that implicitly posit rich countries as the unique site of knowledge production and thus as the source of unidirectional knowledge flows. It also provides a concrete example for consideration of the contexts and practices of postcolonial science, its constraints as well as its promise. The failure of iThemba, which closed its doors in 2015, both is and is not the end of the story.

iThemba was in some senses a one-off. It was the instantiation of a particular network, involving a distinct international group of prominent drug discovery scientists, specific South African senior scientists and funders, and an evolving team of talented young South African–trained bench scientists from across the country and the region. Its model was an unusual form of public-private partnership and involved an unusual confluence of the local and the global. iThemba failed in narrow but important pharmaceutical industry senses: it failed to get new drugs to market, much less into bodies, and it failed to achieve financial sustainability, much less turn a profit.[1]

As an incorporated business entity, iThemba failed for multiple reasons. Many of the problems faced by most start-ups—challenges of leadership, management styles, personnel issues, cash flow—played a role in iThemba's failure. But these alone are inadequate as an explanation of the company's trajectory. On one level, the failure doesn't even merit an explanation: most start-up businesses fail; they sell hopes without ever getting to the point of selling the heralded end products. As anthropologist Kaushik Sunder Rajan has argued, in the space of biotech—with its characteristically long development pipelines—"a significant proportion of the lives and histories" of biotech companies "are stories of these companies having to sell visions of their future products as much as or more than selling the products themselves."[2] iThemba was ultimately unable to keep selling its vision of a future.

Certainly, the proximate cause of iThemba's failure was the South African Technology Innovation Agency's decision that it would not continue to give iThemba the funds that were essential for its operation. But that leaves unanswered the question of why that financial support ended. One reason for that governmental change of heart, which I touched on in earlier discussion of hopes for a knowledge economy, was the desire on the part of some South African officials for a faster return on investment than any drug discovery enterprise could provide: they would rather fund information technology start-ups that can produce a product, and generate return on investment, in the space of months rather than years. This preference is connected to a more fundamental belief held by some government officials and other stakeholders with whom I had spoken from time to time over the years of my research: they think that pharmaceutical R&D is simply too expensive and uncertain for a developing country like South Africa and should be left to the rich countries.

Ultimately, the South African government disinvested in what anthropologist Mike Fortun would characterize as the "story stocks" of iThemba. There are parallels between iThemba and the bankrupted Icelandic genomic technology company at the core of Fortun's ethnography, in which the value of genetic information is "far more promised than present—a value built on promising 'platform technologies,' promising drug targets, patents on promising gene fragments, and the future work of a talented, promising scientific staff."[3] The South African government funders were no longer willing to buy into iThemba's story, and that effectively brought the incorporated entity to its end.

I was drawn to this project because I found iThemba's aspirations for discovering innovative drugs by and for Africans to be compelling, and over the years I developed a deep sympathy with the idealistic and dedicated scientists involved. Despite full awareness of the long odds that they faced, their enthusiasm was infectious. The failure of iThemba is sobering, even as it came to seem overdetermined. Big dreams are not enough to make companies run. Navigating the challenges of funding, markets, global competition, trade deals, intellectual property regimes, access to material—it's daunting. After hope fails, then what?

However, in broader senses, iThemba was arguably *not* a failure. iThemba's closure might be the end of one instantiation of pharmaceutical hope in South Africa, but it is not the end of the enterprise as a whole. Pharmaceutical knowledge-making hopes continue, and the scientists who participated in iThemba now participate in other projects toward those ends.

Moreover, if we consider iThemba itself to be an experiment in the endeavor of creating drug discovery capacity in South Africa, it might be understood as an example of what sociologists Noortje Marres and Linsey McGoey have characterized as *generative failure*—"an occasion for the articulation of issues and mobilization of actors," because it "provides a moment in which situations, objects or relations are rendered legible, and may become subject to inquiry, debate or intervention."[4] Through the failure of iThemba, we can gain a new perspective on the configuration of the worlds of global pharmaceutical science and post-apartheid South Africa in a particular postcolonial moment. The failure of iThemba gives us the opportunity to understand and analyze different key actors of different scales and to look into their attributes and how they interact with each other—ranging from the reagents imported for drug discovery experiments, to the space of the postcolonial synthetic chemistry lab, to the landscape of an economy dominated by extraction industries, to South Africa's place in global trade agreements and scientific exchanges.

For now, this articulation provided by iThemba's failure is discouraging: the Global North retains its centrality in pharmaceutical knowledge making; postcolonial countries compete with each other for an economically viable and epistemically respected niche; research with limited global prestige is what remains palatable for funding in Africa. This small company on the edge of the grounds of the historic Nobel dynamite factory was unable to emerge fully out of the shadows of extraction-based industrialization into the shining future of a knowledge economy: South Africa remains dominated by extraction industries, and aspirations to leave that behind and move into a greener, healthier, and more just future remain frustrated. Yet even without achieving realignment of the global pharmaceutical industry or of South Africa, the iThemba project can illuminate that industry and country, and how they might be otherwise.

Afterlives of Scientific Capacity Building

The hope enacted at iThemba was an active process, not just idle dreams. It is a hope that has material and semiotic consequences. For both South African scientific capacity and for scholarship of global health and post-colonial science, the hope has afterlives.

The site of iThemba feels like a ruin. As of my last visit in June 2017, the building sat with its "iThemba" sign falling down, vacant of people but still full of chemicals and now-idle machines. This equipment was a considerable investment, but there seems to have been insufficient clarity among the South African Technology Innovation Agency funders about who should decide where it goes next. The scientific materials, alas, appear to have been abandoned—but the scientific skills live on.

Even though iThemba was unsuccessful in its mission of drug discovery, the company did make significant progress toward the complementary goal of building scientific capacity in South Africa. For iThemba's cofounders, both South African and international, this capacity building beyond the traditional geographic enclaves of pharmaceutical science was perhaps the most central intervention that they hoped that the iThemba project would make into the realignment of the South African economy and the global pharmaceutical industry. The paths that iThemba scientists took after the company's failure show that this capacity building has had diverse ongoing impacts.

Several of the bench scientists involved have found a new home in another South African drug discovery enterprise: H3D, a public-private drug discovery and development initiative based at the University of Cape Town that was founded in 2011.[5] Indeed, the relationship between those scientists and H3D started even before iThemba closed. Groups of scientists from the two companies had collaborated as part of their participation in major global initiatives such as the Drugs for Neglected Diseases Initiative, and some iThemba scientists moved to H3D well before the former failed. In a sense, iThemba served as an incubator for scientific talent that would contribute to the pool from which H3D could draw.

The work of the former iThemba scientists now at H3D has been continuous to a significant degree, since H3D's mission resonates with iThemba's, as it seeks to "discover and develop innovative, lifesaving medicines for African patients."[6] There are some notable differences. H3D claims to be "Africa's first integrated drug discovery and development centre," and that small word "integrated" points to one of the advantages that it has relative to its predecessor iThemba. H3D has a larger staff of scientists, including not only chemists (like iThemba) but also biologists and more, and thus has the capacity to research promising compounds much further in-house. Its charismatic leader, Kelly Chibale, is effective at addressing both domestic and international audiences.[7] Because it is situated at a university rather than being a for-profit company, it is a better candidate for philanthropic money (including Gates Foundation funds, which had been elusive for iThemba but which H3D has managed to secure). Certainly, H3D faces challenges, but as of this writing it has managed to last longer than iThemba did and to make significant progress, both in building up a staff and funding model set up for success and in promising research on malaria.[8]

Many former iThemba scientists have continued their pharmaceutical knowledge-making endeavors at South African academic institutions. In addition to those at H3D at the University of Cape Town, others are now variously situated at the University of Pretoria, at Rhodes University in Grahamstown, at the University of Johannesburg, and at the University of the Witwatersrand. As an afterlife of iThemba's work, the most notable of these endeavors is the flow chemistry effort at the University of Pretoria.[9] The iThemba scientist who had been sent for training in flow chemistry at the University of Cambridge has now managed to import the needed machinery and has been building an impressive team. Many of the scientists describe their transitions into academia as culturally challenging, and frustratingly slower paced, but they seem happy. And they are training students in the skills that could find a home in future South African drug discovery ventures, informed by a widened scope of scientific knowledge making beyond extraction industries.

These students in training and scientists in development could well find their way to promising biosciences efforts in South Africa that are less directly related to iThemba. For example, the biosciences efforts at the government-run Council for Scientific and Industrial Research have made considerable progress, especially in leads on malaria. They have managed to set up a lab in Pretoria that allows the study of the full life cycle of the malaria parasite at a level matched in only a handful of labs worldwide.

In addition to the specifically scientific knowledge-making element, the experience of working at this start-up pharmaceutical company has also provided scientists with business experiences that they take with them in their ongoing careers. A couple of the scientists have set up their own small consulting companies, the projects of which are a mix of directly related projects like chemical and pharmaceutical importing and formulation, related projects like specialized chemistry services, and further-afield projects such as recruitment. They reflect on their time at iThemba as formative in helping them to understand the challenges and opportunities of doing business in South Africa.

The Scientific Advisory Board members mostly continue as before, even as some are nearing retirement—notably Steve Ley at Cambridge and Tony Barrett at Imperial College London. Cofounder Dennis Liotta is focusing on different collaborations, especially DRIVE (Drug Innovation Ventures at Emory), which is linked with the Emory Institute for Drug Development.[10] That institute was originally founded as the Emory Institute for Drug Discovery, but Liotta has become increasingly focused on development as the key stage at which promising drugs get lost. Liotta has brought one of his longtime collaborators from iThemba and well before, George Painter, to Emory to work on the project. The African Explosives and Chemical Industries leader who had set up the labs, Frank Fisher, is finally fully retired in the rural fly-fishing town of Dullstroom, Mpumalanga.

Although iThemba never found new drugs for TB, HIV, and malaria, it achieved the founders' complementary goal of building up scientific capacity in South Africa. It did not contribute to brain drain. At the time of this writing, only one of the key players on the ground at iThemba has left South Africa. The chief scientific officer, who was originally from the United Kingdom, went to Ireland to become the director of the Pharmaceutical Manufacturing Technology Centre at the University of Limerick and then on to Teva Pharmaceuticals' pharmaceutical development enterprise there.[11] Thus, almost all the human resources that iThemba fostered in South Africa remain in South Africa.

Even if iThemba as a particular instantiation of hope in drug discovery in South Africa was a business failure, hope for a South African role in contributing to such efforts has endured.

Postcolonial Knowledge Making: A Tragedy

iThemba scientists can provide a perspective on living in a particularly productive tension with regard to place. Notions of "putting knowledge in its place" have been fundamental to postcolonial inquiry into science and technology,[12] and "place-based approaches" have gained renewed salience as a counterpoint to "the global."[13] Place figures much less explicitly in laboratory sciences than it does in the field sciences or in clinical practice, but laboratory sciences have long been the more privileged sites of knowledge making.[14] We have much to learn from the laboratories of the postcolonies, not just the metaphorical "laboratories of modernity"[15] but those that have things like beakers and target molecules and nuclear magnetic resonance imaging machines. As synthetic chemists trained in and working with globally standard methods and materials, iThemba scientists' endeavors were not African in any ethnoscience sense. At the same time, this particular locality in science not only is interesting in its own right but also offers a valuable vantage point from which to think through synthetic chemistry as an endeavor that necessarily takes place in particular locations with particular material practices even as its participants contribute to a globally mobile scientific field. These scientists living and working in South Africa acknowledge and accept place while also seeking lines of flight by which it might be transcended.

The unachieved transcendence of place points to a second valence of "putting knowledge in its place" that can help us to understand iThemba and its failure. To *put a person in their place* is to remind someone of their position, which is to say, to humble them. If those involved in iThemba thought that they could dream big, the company's failure was a rude awakening. In this sense, iThemba's story takes the form of a tragedy: the small company might be understood to play the role of the hero who was perhaps destined from the beginning to fail in the face of overpowering forces—in this case, South African and global social, economic, and political orders.

In the trope of tragedy, the qualities that have led to the hero's rise are often the same qualities that lead to the hero's ultimate downfall. Reading iThemba's emergence and failure in the trope of tragedy can help to illuminate the broader landscape of the potential for drug discovery capacity in South Africa. The same elements that created the conditions of possibility for building drug discovery capacity in South Africa also became powerful constraints: the extraction-industry-based industrialization of South Africa made pharmaceutical manufacturing there conceivable while also crowding it out; the imperative to serve the needs of a democratic post-apartheid South Africa made the mission of creating pharmaceuticals by and for South Africa compelling and yet also rendered it precariously subject to public funders' short timelines; international interest in contributing to South Africa's success allowed the mobilization of energy and expertise among elite international drug discovery scientists at the same time that global health discourses overwhelmingly marginalize African scientists because Africans are framed as objects of, rather than as contributors to, global knowledge.

The story of iThemba is one of the place and matter of pharmaceutical knowledge making, as it is and as it might be. Paying attention to iThemba in this way helps to reveal the material and informational infrastructures that fostered and impeded it. Engagements with matter have become popular in science and technology studies and more broadly, and this synthetic-chemistry-based drug discovery effort offers a distinctive opportunity to engage with this question. I have followed matter across scales, attentive to the lab as part of infrastructural networks. As anthropologist Brian Larkin observes, because "infrastructures are matter that enable the movement of other matter," they have a "peculiar ontology" as "things and also the relation between things."[16] On a material level, these efforts to build drug discovery science sought to both create things and to intervene in how things become mobile. On a semiotic level, this mobilization of infrastructure in new ways would also restructure relations between groups of people. Ethnographer of infrastructure Susan Leigh Star has pointed out that "infrastructure does not grow *de novo*; it wrestles with the inertia of the installed base and inherits strengths and limitations from that base."[17] As I have shown, the infrastructures of mining and post-apartheid create the contours of the strengths and limitations of South African drug discovery.

iThemba can be understood as an intervention into the configuration of the place of South Africa in the world. South Africa in my account has been articulated as a simultaneously literal and imaginary place, and the effort to build drug discovery capacity there is also simultaneously a practical endeavor and a sociotechnical imaginary. The desired transformation of the map of drug discovery was ultimately insufficiently realized: the metropoles of pharmaceutical science maintain their centrality, and the competition among postcolonial countries to join that enterprise does not alter that. HIV, TB, and malaria remain serious problems for South Africa and its region. And yet iThemba's endeavor provides an alternative vision for how new treatments might be developed and for what South Africa could (yet) be.

The tragedy does not repudiate the heroism of the protagonist's journey. This particular effort to transform the treatment of HIV, TB, and malaria while repositioning South Africa in the landscape of global science was unsuccessful, but the failure of the instance is not the failure of the dream. Global capital inevitably spurs creative entrepreneurialism in peripheral places, and injustices and inequalities inevitably spur resistance and efforts for change. The specific future forms will always be unknown, but in some form or other, there will always be new hope.

ACKNOWLEDGMENTS

First and foremost, my gratitude goes to all the scientists involved with iThemba Pharmaceuticals, who were so generous with their time and insights.

This book benefited greatly from feedback on presentations of this work in progress, including multiple Society for Social Studies of Science meetings and meetings of the National Women's Studies Association, Society for the History of Technology, International Congress of History of Science and Technology, American Anthropological Association, and African Studies Association. Smaller conferences and workshops were particularly fruitful sites for developing these ideas: the "Lives of Property" conference at Oxford University, the "Biomedical Objects in Africa" workshop at Johns Hopkins, the "Innovation, Transformation and Sustainable Futures in Africa" conference in Dakar, the "Dreaming of Health and Science in Africa" conference in Cambridge, United Kingdom, and the "Africanizing Technology" conference at Wesleyan University. I also benefited from invitations to speak at the Indian Institute of Technology–Delhi, Sydney University, the University of the Witwatersrand, Emory University, the University of Pretoria, and especially at forums in which earlier drafts were precirculated: a Johns Hopkins History of Science, Technology, and Medicine Colloquium; a "Feminist Postcolonial STS" workshop at the University of Michigan; and a Culture, Medicine, and Power Conversation at the Department of Social Science, Health, and Medicine at King's College London.

My beloved writing group at Georgia Tech was a critical source of both concrete feedback and enduring support throughout: Nassim JafariNaimi, Mary McDonald, and Jennifer Singh. Being in community with these women has been an essential part of my process. We faced the ups and downs of academic writing cycles together, challenging and sustaining each other by being critically engaged yet generous readers—and real friends.

More broadly, Georgia Tech provided intellectual community and institutional support, and I am especially grateful to Wenda Bauchspies, Jennifer Clark, Carol Colatrella, Susan Cozzens, Shatakshee Dhongde, John Krige, Janet Murray, Manu Platt, and Lewis Wheaton, as well as Renee Shelby and my research assistant Rebecca Watts Hull.

My new academic home, King's College London, was an excellent place at which to complete the book's final stretch. The book had already been shaped by vital conversations with the Culture, Medicine and Power Research Group there, and the move has allowed me to deepen that engagement. I also benefited from the support of Bronwyn Parry, Karen Glaser, Nikolas Rose, and Susan Chandler.

Academics based in South Africa were also vital interlocutors over the years, including Rachel Cesar, Kelly Gillespie, Julia Hornberger, Nolwazi Mkhwanazi, and especially David Walwyn and Catherine Burns. I have also had the privilege to be in conversation with many other scholars who gave me feedback along the way, including Amy Agigian, Ruha Benjamin, Melinda Cooper, Melissa Creary, Joe Dumit, Laura Foster, Jeremy Greene, Ann Kelly, Linsey McGoey, Alondra Nelson, Tolu Odumosu, Kane Race, Marsha Rosengarten, Deboleena Roy, Miriam Shakow, Lindsay Smith, Banu Subramaniam, Kim TallBear, Natali Valdez, and Scott Vrecko.

I also thank others who contributed time and talents: Katherine Behar provided photographs, Hannah Dar provided an illustration, and Holly O'Neill provided editing assistance.

At the University of Chicago Press, Karen Darling's support for the project has been instrumental, as were the comments from the extraordinarily engaged and helpful anonymous peer reviewers.

The research for this book received financial support from Georgia Tech, including start-up funds and the Ivan Allen College Dean's Small Grants for Research, from the National Science Foundation (Award 1331049), and from King's College London. I published an early iteration of the ideas that would become developed in this book as "Places of Pharmaceutical Knowledge-Making: Global Health, Postcolonial Science, and South African Drug Discovery," *Social Studies of Science* 44, no. 6 (December 2014): 848–73.

Maital Dar has provided invaluable support at every step of this book's journey, accompanying me on that first wintery trip to Johannesburg and through innumerable iterations of writing and editing since. I cannot possibly thank her enough.

NOTES

INTRODUCTION

1. Jean Comaroff and John L. Comaroff, *Theory from the South; or, How Euro-America Is Evolving toward Africa* (Boulder: Paradigm, 2011), 1.
2. Sjaak van der Geest, "Anthropology and the Pharmaceutical Nexus," *Anthropological Quarterly* 79, no. 2 (2006): 312.
3. Anthropological critique of global health paradigms rooted in ethnographic research in Africa has been extraordinarily rich, as highlighted in Andrew McDowell, "Making the Global Equivalent: Markets, Relations and Pharmaceuticals in the Anthropology of Global Health in Africa," *BioSocieties* 10, no. 3 (September 2015): 380–84.
4. Kristin Peterson, "AIDS Policies for Markets and Warriors: Dispossession, Capital, and Pharmaceuticals in Nigeria," in *Lively Capital: Biotechnologies, Ethics, and Governance in Global Markets*, ed. Kaushik Sunder Rajan (Durham, NC: Duke University Press, 2012), 228–47. At the time that it closed its doors in 2015, iThemba Pharmaceuticals had just one rival in South Africa, a company called H3D, which is based at the University of Cape Town campus. That company also relies on international partnerships but on a somewhat different public-private model. See Linda Nordling, "Made in Africa," *Nature Medicine* 19, no. 7 (2013): 803–6. As of this writing in 2018, H3D is the only drug discovery company in South Africa.
5. There are relatively few exceptions to the more general lack of anthropological rapport with employees of pharmaceutical companies, as pointed out in Van der Geest, "Anthropology and the Pharmaceutical Nexus," esp. 312.
6. The definition of "Big Pharma" is imprecise but includes such market-leading companies as Johnson & Johnson (2015 market value = $276 billion), Novartis ($273 billion), Roche ($248 billion), Pfizer ($212 billion), and Merck ($164 billion). See Sean Williams, "7 Facts You Probably Don't Know about Big Pharma," *Motley Fool*, July 19, 2015, accessed December 26, 2017, https://www.fool.com/investing/value/2015/07/19/7-facts-you-probably-dont-know-about-big-pharma.aspx.
7. There were African sites that played important roles in pharmaceutical research in the colonial era, when research and therapy were more intimately connected. See, e.g., Deborah Neill, "Paul Ehrlich's Colonial Connections: Scientific Networks and

Sleeping Sickness Drug Therapy Research, 1900–1914," *Social History of Medicine* 22, no. 1 (2009): 61–77. However, that is no longer the case.

8. David Wade Chambers and Richard Gillespie, "Locality in the History of Science: Colonial Science, Technoscience, and Indigenous Knowledge," *Osiris* 15 (2000): 223.
9. Paul N. Edwards and Gabrielle Hecht, "History and the Technopolitics of Identity: The Case of Apartheid South Africa," *Journal of Southern African Studies* 36, no. 3 (September 2010): 619–39.
10. Barriers to true participant observation are common in elite technoscientific settings, as discussed by Hugh Gusterson, "Studying Up Revisited," *PoLAR: Political and Legal Anthropology Review* 20, no. 1 (1997): 116.
11. For Liotta's account of this history, see Dennis C. Liotta and George R. Painter, "Discovery and Development of the Anti–human Immunodeficiency Virus Drug, Emtricitabine (Emtriva, FTC)," *Accounts of Chemical Research* 49, no. 10 (2016): 2091–98.
12. Laura Nader, "Up the Anthropologist: Perspectives Gained from Studying Up," in *Revisiting Anthropology*, ed. Dell Hymes (1969; New York: Vintage Books, 1974), 284–311.
13. See Kim TallBear's discussion of her engagement with the increasing number of genetic scientists who are women and people of color: *Native American DNA: Tribal Belonging and the False Promise of Genetic Science* (Minneapolis: University of Minnesota Press, 2013), 18. For an insightful exploration of positionality of northern social researchers studying southern scientific elites, see Logan D. A. Williams, "Mapping Superpositionality in Global Ethnography," *Science, Technology, and Human Values* 43, no. 2 (2018): 198–223.
14. Chris D. Edlin et al., "Identification and In-Vitro ADME Assessment of a Series of Novel Anti-malarial Agents Suitable for Hit-to-Lead Chemistry," *American Chemical Society Medicinal Chemistry Letters* 3, no. 7 (July 12, 2012): 570–73.
15. For discussion of high-throughput versus knowledge-based approaches, see J. P. Hughs et al., "Principles of Early Drug Discovery," *British Journal of Pharmacology* 162 (2011): 1242.
16. Chris Edlin, "The Importance of Patent Sharing in Neglected Disease Drug Discovery," *Future Medicinal Chemistry* 3, no. 11 (August 31, 2011): 1331–34.
17. M. S. Sonopo et al., "Carbon-14 Radiolabeling and In Vivo Biodistribution of a Potential Anti-TB Compound," *Journal of Labelled Compounds and Radiopharmaceuticals* 58, no. 2 (February 2015): 23–29.
18. Although designed in a rich country, the page does not give the impression of being designed by a professional design firm because of its dark-red and dusty-blue color choice, multiple typefaces, and imperfectly spaced text.
19. See Achille Mbembe, *On the Postcolony* (Berkeley: University of California Press, 2001).
20. See Kelly Gillespie, "Reclaiming Nonracialism: Reading the Threat of Race from South Africa," *Patterns of Prejudice* 44, no. 1 (2010): 73.
21. James Ferguson, *Global Shadows: Africa in the Neoliberal World* (Durham, NC: Duke University Press, 2006); see esp. 6.
22. Erik Saethre and Jonathan Stadler, *Negotiating Pharmaceutical Uncertainty: Women's Agency in a South African HIV Prevention Trial* (Nashville, TN: Vanderbilt University Press, 2017).
23. Jean B. Nachega et al., "Lower Pill Burden and Once-Daily Dosing Antiretroviral

Treatment Regimens for HIV Infection: A Meta-analysis of Randomized Controlled Trials," *Clinical Infectious Diseases* 58, no. 9 (2014): 1297–1307.

24. P. Wenzel Geissler, "Public Secrets in Public Health: Knowing Not to Know When Making Scientific Knowledge," *American Ethnologist* 40, no. 1 (2013): 19.
25. Richard Milne, "Pharmaceutical Prospects: Biopharming and the Geography of Technological Expectations," *Social Studies of Science* 42, no. 2 (2012): 299.
26. Some of the most evocative stories of how scarcity shapes clinical practice are in relatively developed African contexts, such as Botswana and urban Nigeria. For the former, see Julie Livingston, *Improvising Medicine: An African Oncology Ward in an Emerging Epidemic* (Durham, NC: Duke University Press, 2012); for the latter, see Iruka N. Okeke, *Divining without Seeds: The Case for Strengthening Laboratory Medicine in Africa* (Ithaca, NY: Cornell University Press, 2011).
27. Anna Lowenhaupt Tsing, *Friction: An Ethnography of Global Connection* (Princeton, NJ: Princeton University Press, 2005), 1.
28. Chambers and Gillespie, "Locality in the History of Science," 229.
29. Megan Bradley, "On the Agenda: North-South Research Partnerships and Agenda-Setting Practices," *Development in Practice* 18, no. 6 (November 2008): 674–75.
30. Thomas C. Nchinda, "Research Capacity Strengthening in the South," *Social Science and Medicine* 54, no. 11 (June 2002): 1708.
31. Saul A. Dubow, *A Commonwealth of Knowledge: Science, Sensibility and White South Africa, 1820–2000* (Oxford: Oxford University Press, 2006).
32. Sheila Jasanoff and Sang-Hyun Kim, "Containing the Atom: Sociotechnical Imaginaries and Nuclear Power in the United States and South Korea," *Minerva* 47, no. 2 (2009): 123.
33. Sheila Jasanoff, "Future Imperfect: Science, Technology, and the Imaginations of Modernity," in *Dreamscapes of Modernity: Sociotechnical Imaginaries and the Fabrication of Power*, ed. Sheila Jasanoff and Sang-Hyun Kim (Chicago: University of Chicago Press, 2015), 4.
34. Ibid.
35. Ibid., 25.
36. Ibid., 19.
37. Edwards and Hecht, "History and the Technopolitics of Identity."
38. Stephen John Sparks, "Apartheid Modern: South Africa's Oil from Coal Project and the History of a South African Company Town" (PhD diss., University of Michigan, 2012).
39. Keith Breckenridge, *Biometric State: The Global Politics of Identification and Surveillance in South Africa, 1850 to the Present* (Cambridge: Cambridge University Press, 2014).
40. This is an idea that I developed in my first book, *Medicating Race: Heart Disease and Durable Preoccupations with Difference* (Durham, NC: Duke University Press, 2012); see esp. 19.
41. Joseph Dumit, *Drugs for Life: How Pharmaceutical Companies Define Our Health* (Durham, NC: Duke University Press, 2012); Jeremy A. Greene, *Prescribing by Numbers: Drugs and the Definition of Disease* (Baltimore: Johns Hopkins University Press, 2007).
42. Kane Race, *Pleasure Consuming Medicine: The Queer Politics of Drugs* (Durham, NC: Duke University Press, 2009).
43. Sjaak van der Geest, Susan Reynolds Whyte, and Anita Hardon, "The Anthropology of Pharmaceuticals: A Biographical Approach," *Annual Review of Anthropology* 25 (1996): 154.

44. Ibid., 154–55.
45. Judith Butler, *Bodies That Matter: On the Discursive Limits of Sex* (New York: Routledge, 1993).
46. Notable exceptions include Elizabeth A. Wilson, "The Work of Antidepressants: Preliminary Notes on How to Build an Alliance between Feminism and Psychopharmacology," *BioSocieties* 1 (2006): 125–31; Asha Persson, "Incorporating Pharmakon: HIV, Medicine, and Body Shape Change," *Body and Society* 10, no. 4 (2004): 45–67; Marsha Rosengarten, *HIV Interventions: Biomedicine and the Traffic between Information and Flesh* (Seattle: University of Washington Press, 2009).
47. Susan Reynolds Whyte, Sjaak van der Geest, and Anita Hardon, *Social Lives of Medicines* (Cambridge: Cambridge University Press, 2002), 37.
48. For a contrast, see Dumit, *Drugs for Life*.
49. João Biehl, "Pharmaceuticalization: AIDS Treatment and Global Health Politics," *Anthropological Quarterly* 80, no. 4 (2007): 1083–126.
50. For India, see Veena Das and Ranendra K. Das, "Pharmaceuticals in Urban Ecologies: The Register of the Local," in *Global Pharmaceuticals: Ethics, Markets, Practices*, ed. Adriana Petryna, Andrew Lakoff, and Arthur Kleinman (Durham, NC: Duke University Press, 2006), 171–205; Stefan Ecks, "Pharmaceutical Citizenship: Antidepressant Marketing and the Promise of Demarginalization in India," *Anthropology and Medicine* 12, no. 3 (December 2005): 239–54. For Mexico, see Cori Hayden, "A Generic Solution? Pharmaceuticals and the Politics of the Similar in Mexico," *Current Anthropology* 48, no. 4 (2007): 475–95. For West Africa, see Vinh-Kim Nguyen, "Anti-retroviral Globalism, Biopolitics, and Therapeutic Citizenship," in *Global Assemblages: Technology, Politics, and Ethics as Anthropological Problems*, ed. Aihwa Ong and Stephen J. Collier (Malden, MA: Blackwell, 2005), 124–44.
51. Audre Lorde, "The Master's Tools Will Never Dismantle the Master's House," in *Sister Outsider: Essays and Speeches* (Trumansburg, NY: Crossing Press, 1984), 110–13. For further discussion of this tension, see Anne Pollock and Banu Subramaniam, "Resisting Power, Retooling Justice: Promises of Feminist Postcolonial Technosciences," *Science, Technology, and Human Values* 41, no. 6 (2016): 951–66.
52. See Melinda Cooper, "On Pharmaceutical Empire: AIDS, Security, and Exorcism," in *Life as Surplus: Biotechnology and Capitalism in the Neoliberal Era* (Seattle: University of Washington Press, 2008), 51–73; Didier Fassin, *When Bodies Remember: Experiences and Politics of AIDS in South Africa* (Berkeley: University of California Press, 2007).
53. Bruno Latour, "Give Me a Laboratory and I Will Raise the World," in *Science Observed: Perspectives on the Social Study of Science*, ed. Karin D. Knorr-Cetina and Michael Mulkay (London: Sage, 1983), 147.
54. Bruno Latour, *Aramis; or, The Love of Technology* (Cambridge, MA: Harvard University Press, 1996), 23, 56–58.
55. Hegel himself did not use precisely these terms, but they are widely associated with Hegelian logic. See David Gray Carlson, *A Commentary on Hegel's Science of Logic* (London: Palgrave Macmillan, 2007), 23n45.
56. See Karl Marx, *The Poverty of Philosophy* (1847), trans. Harry Quelch (New York: Cosimo 2008).
57. Cheryl Mattingly, *The Paradox of Hope: Journeys through a Clinical Borderland* (Berkeley: University of California Press, 2010), 6.
58. Ibid., 3.

59. Kaushik Sunder Rajan, *Biocapital: The Constitution of Postgenomic Life* (Durham, NC: Duke University Press, 2006).
60. Paulo Freire, *Pedagogy of Hope: Reliving "Pedagogy of the Oppressed,"* trans. Robert R. Barr (London: Bloomsbury Academic, 2014), 2.
61. Nathan Greenslit, "Dep®ession and Comsum♀tion: Psychopharmaceuticals, Branding, and New Identity Practices," *Culture, Medicine and Psychiatry* 29, no. 4 (December 2005): 477–502.
62. Dumit, *Drugs for Life*, 73.
63. Transcript of speech by Barack Obama at the Democratic National Convention, July 27, 2004, accessed December 30, 2016, http://www.librarian.net/dnc/speeches/obama.txt.

CHAPTER 1

1. Emory Law's conference "Advancing the Consensus" was held on October 18, 2008, to celebrate the sixtieth anniversary of the Universal Declaration of Human Rights. The panelists for "International Access to Medicines" were Dr. Dennis C. Liotta, professor of organic chemistry, Emory University; Todd Sherer, director, Office of Technology Transfer, Emory University; and Sherry M. Knowles, GlaxoSmithKline.
2. This argument operates on the widely held view that high drug prices in the Global North are due to the high cost of R&D rather than what the market will bear, an idea that I explore elsewhere: Anne Pollock, "Transforming the Critique of Big Pharma," *BioSocieties* 6, no. 1 (2011): 106–18.
3. "2007 Global Health Partnership Program Grants—Republic of South Africa Drug Discovery Training Program," Emory Global Health Institute Global Health Partnership Grants, accessed December 26, 2017, http://www.globalhealth.emory.edu/what/faculty_programs/ghpp_grants/2007_republic_sa.html.
4. Quotation from iThemba's website, accessed August 8, 2013, ithembapharma.com.
5. The issue of disproportionate investment is widely known as the 10/90 gap, in which "less than 10% of this [global R&D expenditure] is devoted to diseases or conditions that account for 90% of the global disease burden." Global Forum for Health Research, *The 10/90 Report on Health Research 2000* (Geneva: World Health Organization, 2000), accessed December 26, 2017, http://announcementsfiles.cohred.org/gfhr_pub/assoc/s14791e/s14791e.pdf.
6. Amy Laura Hall, "Whose Progress? The Language of Global Health," *Journal of Medicine and Philosophy* 31 (2006): 289. In South Africa in particular, there is a long intertwining of colonialism and biomedicine: Jean Comaroff, "The Diseased Heart of Africa: Medicine, Colonialism, and the Black Body," in *Knowledge, Power, and Practice: The Anthropology of Medicine and Everyday Life*, ed. Shirley Lindenbaum and Margaret Lock (Berkeley: University of California Press, 1993), 305–29.
7. See http://essentialmedicine.org.
8. Universities Allied for Essential Medicines, "2013 Annual Conference Guide," 1, accessed November 30, 2015, http://uaem.org/cms/assets/uploads/2013/09/2013UAEMConference_Packet.pdf.
9. The sense of responsibility toward the Other is of course preferable to disregard for the Other, but as Nelson Maldonado-Torres has so persuasively argued, the *ego cogito* (of Descartes's "I think therefore I am") is built on the foundation of the conquering self, or *ego conquiro*, and it maintains skepticism about whether the colonized Other can think. Nelson Maldonado-Torres, "On the Coloniality of Being:

Contributions to the Development of a Concept," *Cultural Studies* 21, nos. 2–3 (2007): esp. 245: "I am suggesting that the practical conquering self and the theoretical thinking substance are parallel in terms of their certainty. The *ego conquiro* is not questioned, but rather provides the ground for the articulation of the *ego cogito*."

10. For further discussion, see Pollock "Transforming the Critique of Big Pharma."
11. Steven Robins, "'Long Live Zackie, Long Live': AIDS, Activism, Science and Citizenship after Apartheid," *Journal of Southern African Studies* 30, no. 3 (September 2004): 651–72. See also Nicoli Nattrass, *Mortal Combat: AIDS Denialism and the Struggle for Antiretrovirals in South Africa* (Scottsville: University of KwaZulu-Natal Press, 2007).
12. Jean Comaroff, "Beyond Bare Life: AIDS, (Bio)politics, and the Neoliberal Order," *Public Culture* 19, no. 1 (December 2007): 211.
13. See Jeremy A. Greene, *Generic: The Unbranding of Modern Medicine* (Baltimore: Johns Hopkins University Press, 2014); Hayden, "Generic Solution?"; Cori Hayden, "The Proper Copy: The Insides and Outsides of Domains Made Public," *Journal of Cultural Economy* 3, no. 1 (March 2010): 85–102.
14. Hayden, "Proper Copy," 85.
15. See discussion in Greene, *Generic*; Hayden, "Generic Solution?"; Manaf Kottakkunnummal, "The Social Life of Indian Generic Pharmaceuticals in Johannesburg" (PhD diss., University of the Witwatersrand, 2016), 108.
16. See Greene, *Generic*, 16.
17. Antoinette Burton, *Africa in the Indian Imagination: Race and the Politics of Postcolonial Citation* (Durham, NC: Duke University Press, 2016), 4.
18. Roger Bate, Ginger Zhe Jin, Aparna Mathur, and Amir Attaran, "Poor Quality Drugs in Global Trade: A Pilot Study," NBER Working Paper 20469, September 2014.
19. Aarti Patel et al., "'This Body Does Not Want Free Medicines': South African Consumer Perceptions of Drug Quality," *Health Policy and Planning* 25 (2010): 61–69.
20. Joan Rovira, "Creating and Promoting Domestic Drug Manufacturing Capacities: A Solution for Developing Countries?," in *Negotiating Health: Intellectual Property and Access to Medicines*, ed. Pedro Roffe, Geoff Tansey, and David Vivas-Eugui (London: Earthscan, 2006), 231.
21. Palesa Sekhejane and Charlotte Pelletan, "HIV and AIDS Triumphs and Struggles," in *Sizonqoba! Outliving AIDS in Southern Africa*, ed. Busani Ngcaweni (Pretoria: Africa Institute of South Africa, 2016), 97.
22. Kottakkunnummal, "Social Life of Indian Generic Pharmaceuticals in Johannesburg."
23. Paul Farmer has questioned the tendency in public health literature to call for less funding for high-tech solutions and more for low-tech preventive ones, asking why wasn't it that the "dilemmas of the Haitian sick call for a full range of high tech *and* low tech innovations?" Paul Farmer, *Infections and Inequalities: The Modern Plagues* (Berkeley: University of California Press, 1999), 21.
24. Anne Pollock, "Pharmaceutical Meaning-Making beyond Marketing: Racialized Subjects of Generic Thiazide," *Journal of Law, Medicine, and Ethics* 36, no. 3 (2008): 530–36.
25. Peter Redfield, "Bioexpectations: Life Technologies as Humanitarian Goods," *Public Culture* 24, no. 1 (2012): 157–84, citing Fiona Terry, *Condemned to Repeat? The Paradox of Humanitarian Action* (Ithaca, NY: Cornell University Press, 2002).
26. Farmer, *Infections and Inequalities*, 21.
27. Semiotically laden resistance to being relegated to generic drugs is something that I explored previously with regard to African Americans: see Anne Pollock, "Thiazide

Diuretics at a Nexus of Associations: Racialized, Proven, Old, Cheap," in *Medicating Race*, 131–54.

28. Javier Lezaun and Catherine A. Montgomery, "The Pharmaceutical Commons: Sharing and Exclusion in Global Health Drug Development," *Science, Technology, and Human Values* 40, no. 1 (2015): 5.
29. See Susan L. Erikson, "Secrets from Whom? Following the Money in Global Health Finance," *Current Anthropology* 56, no. S12 (December 2015): S306–S316. For a broader discussion of the complex regimes of knowledge and finance that have led to the ascendance of the PDP model, see Susan Craddock, *Compound Solutions: Pharmaceutical Alternatives for Global Health* (Minneapolis: University of Minnesota Press, 2017).
30. Adriana Petryna, *When Experiments Travel: Clinical Trials and the Global Search for Human Subjects* (Princeton, NJ: Princeton University Press, 2009); Kaushik Sunder Rajan, "Experimental Values: Indian Trials and Surplus Health," *New Left Review* 45 (2007): 67–88.
31. For discussion of these initiatives' models, see Lezaun and Montgomery, "Pharmaceutical Commons."
32. Ibid., 18–19.
33. Albertina Torsoli and Mikiko Kitamura, "Drug Makers Join Gates Foundation to Halt Tropical Illness," Bloomberg, January 30, 2012, accessed December 26, 2012, http://www.bloomberg.com/news/2012-01-30/drugmakers-join-gates-foundation-in-fighting-tropical-diseases.html.
34. Grand Challenges, accessed January 1, 2016, http://grandchallenges.org.
35. Grand Challenges, accessed January 30, 2013, http://www.grandchallenges.org/Pages/GrantsMap.aspx.
36. Noémi Tousignant, "Broken Tempos: Of Means and Memory in a Senegalese University Laboratory," *Social Studies of Science* 43, no. 5 (2013): 747.
37. "Carlos Slim Helu & Family," Profile, *Forbes*, https://www.forbes.com/profile/carlos-slim-helu/.
38. Linsey McGoey, *No Such Thing as a Free Gift: The Gates Foundation and the Price of Philanthropy* (London: Verso, 2015), 9.
39. Between 2009 and 2012, the South African Department of Science and Technology allocated 14 percent of its total budget to the "prestige project" of the Square Kilometre Array, more than double its allocation to the National Research Foundation, which funds doctoral fellowships and basic research. See Michael Cherry, "South African Science: Black, White, and Grey," *Nature* 463 (2010): 727. Physics can, of course, also be articulated as an important postcolonial science project; for a South African discussion of this, see Nithaya Chetty and Ahmed C. Bawa, "Physics for Development in Africa," *APS News* 14, no. 10 (November 2005): 8.
40. David Walwyn, "Determining Quantitative Targets for Public Funding of Tuberculosis Research and Development," *Health Research Policy and Systems* 11, no. 10 (2013): 1–8.
41. Hugh Gusterson's deconstruction of arguments about nuclear weapons in India and China is resonant: "Nuclear Weapons and the Other in the Western Imagination," *Cultural Anthropology* 14, no. 1 (February 1999): 111–43.
42. Cori Hayden, *When Nature Goes Public: The Making and Unmaking of Bioprospecting in Mexico* (Princeton, NJ: Princeton University Press, 2003); Stacey Langwick, *Bodies, Politics, and African Healing: The Matter of Maladies in Tanzania* (Bloomington: Indiana University Press, 2011); Rachel Wynberg, Doris Schroeder, and Roger Chennells,

eds., *Indigenous Peoples, Consent and Benefit Sharing: Lessons from the San-Hoodia Case* (London: Springer, 2009).

43. Markku Hokkanen, "Imperial Networks, Colonial Bioprospecting and Burroughs Wellcome & Co.: The Case of Strophanthus Kombe from Malawi (1859–1915)," *Social History of Medicine* 25, no. 3 (2012): 589–607; Abena Osseo-Asare, "Bioprospecting and Resistance: Transforming Poisoned Arrows into Strophanthin Pills in Colonial Gold Coast, 1885–1922," *Social History of Medicine* 21, no. 2 (2008): 269–90.
44. Laura A. Foster, *Reinventing Hoodia: Peoples, Plants, and Patents in South Africa* (Seattle: University of Washington Press, 2017); Abena Osseo-Asare, *Bitter Roots: The Search for Healing Plants in Africa* (Chicago: University of Chicago Press, 2014).
45. Hanspeter C. W. Reihling, "Bioprospecting the African Renaissance: The New Value of *Muthi* in South Africa," *Journal of Ethnobiology and Ethnomedicine* 4, no. 9 (2008): accessed October 10, 2017, http://www.ethnobiomed.com/content/4/1/9; Adam Ashforth, "Muthi, Medicine and Witchcraft: Regulating 'African Science' in Post-apartheid South Africa?," *Social Dynamics* 31, no. 2 (2005): 211–42.
46. Sandra Harding, ed., *The Postcolonial Science and Technology Studies Reader* (Durham, NC: Duke University Press, 2011).
47. Joshua Rosenthal, "Deconstructing a Research Project," review of *When Nature Goes Public*, by Cori Hayden, *EMBO Reports* 5, no. 11 (2004): 136.
48. Osseo-Asare, *Bitter Roots*.
49. Langwick, *Bodies, Politics, and African Healing*. See also Christopher Morris, "Biopolitics and Boundary Work in South Africa's Sutherlandia Clinical Trial," *Medical Anthropology* 36, no. 7 (2017): 685–98, esp. 694–95.
50. Damien Droney, "Scientific Capacity Building and the Ontologies of Herbal Medicine in Ghana," *Canadian Journal of African Studies* 50, no. 3 (2016): 438.
51. P. Wenzel Geissler and Ruth Prince, "Active Compounds and Atoms of Society: Plants, Bodies, Minds, and Cultures in the Work of Kenyan Ethnobotanical Knowledge," *Social Studies of Science* 39, no. 4 (2009): 615.
52. Hayden, *When Nature Goes Public*; Langwick, *Bodies, Politics, and African Healing*.
53. This is a case analyzed by Laura A. Foster in "Inventing Hoodia: Vulnerabilities and Epistemic Citizenship in Southern Africa," *CSW Update*, April 2011, 15–19, and in "Decolonizing Patent Law: Postcolonial Technoscience and Indigenous Knowledge in South Africa," *Feminist Formations* 28, no. 3 (Winter 2016): 148–73, as well as by Osseo-Asare in *Bitter Roots*.
54. Wynberg, Schroeder, and Chennells, *Indigenous Peoples, Consent and Benefit Sharing*.
55. Helen Tilley, *Africa as a Living Laboratory: Empire, Development, and the Problem of Scientific Knowledge, 1870–1950* (Chicago: University of Chicago Press, 2011).
56. William C. Sturtevant, "Studies in Ethnoscience," *American Anthropologist* 66, no. 3 (June 1964): 99–131.
57. Helen Watson-Verran and David Turnbull, "Science and Other Knowledge Systems," in *Handbook of Science and Technology Studies*, ed. Sheila Jasanoff et al. (London: Sage, 1995), 115–39.
58. See Warwick Anderson, "From Subjugated Knowledge to Conjugated Subjects: Science and Globalisation, or Postcolonial Studies of Science?," *Postcolonial Studies* 12, no. 4 (2009): 389.
59. Ruha Benjamin, "A Lab of Their Own: Genomic Sovereignty as Postcolonial Science Policy," *Policy and Society* 28 (2009): 341–55. See also Banu Subramaniam, "Colonial Legacies, Postcolonial Biologies: Gender and the Promises of Biotechnology," *Asian Biotechnology and Development Review* 17, no. 1 (March 2015): 15–36.

60. Ann H. Kelly and P. Wenzel Geissler, eds., *The Value of Transnational Medical Research: Labour, Participation, and Care* (London: Routledge, 2012); P. Wenzel Geissler and Catherine Molyneaux, *Evidence, Ethos, and Experiment: The Anthropology and History of Medical Research in Africa* (Oxford: Berghahn Books, 2011); Petryna, *When Experiments Travel*; Rajan, "Experimental Values."
61. Richard Rottenburg, "Social and Public Experiments and New Figurations of Science and Politics in Postcolonial Africa," *Postcolonial Studies* 12, no. 4 (2009): 423–40. Of course, African research subjects are not the passive objects that their portrayal in this literature often suggests: see Catherine M. Montgomery, "Making Prevention Public: The Co-production of Gender and Technology in HIV Prevention Research," *Social Studies of Science* 42, no. 6 (December 2012): 922–44.
62. P. Wenzel Geissler, ed., *Para-states and Medical Science: Making African Global Health* (Durham, NC: Duke University Press, 2015), 20.
63. Ferdinand Moyi Okwaro and P. W. Geissler, "In/dependent Collaborations: Perceptions and Experiences of African Scientists in Transnational HIV Research," *Medical Anthropology Quarterly* 29, no. 4 (December 2015): 499.
64. Karen Poltis Virk, "South Africa Today: Economical Development and Regulatory Standards Make the Country a Logical Choice for Trials," *Applied Clinical Trials*, November 1, 2009, accessed October 5, 2017, http://www.appliedclinicaltrialsonline.com/south-africa-today.
65. Mbembe, *On the Postcolony*, 2.
66. Saethre and Stadler, *Negotiating Pharmaceutical Uncertainty*.
67. See Claire L. Wendland, "Research, Therapy, and Bioethical Hegemony: The Controversy over Perinatal AZT Trials in Africa," *African Studies Review* 51, no. 3 (December 2008): 1–23.
68. Selidji T. Agnandji et al., "Patterns of Biomedical Science Production in a Sub-Saharan Research Center," *BMC Medical Ethics* 13, no. 3 (March 2012): 6. Of course, the notion that data can be unmediated by the people and processes of their collection is a fallacy, but it is one that does important ideological work: see Crystal Biruk, *Cooking Data: Culture and Politics in an African Research World* (Durham, NC: Duke University Press, 2018).
69. Although Claire Wendland is focused on global health partnerships for health care provision rather than for clinical trial research, her call to "open up the black box" on "the specificities, histories, and inner workings of African institutions and individuals" in order to develop a more capacious conceptualization of global health capacity is relevant for clinical trial research as well. Claire L. Wendland, "Opening Up the Black Box: Looking for a More Capacious Version of Capacity in Global Health Partnerships," *Canadian Journal of African Studies / Revue canadienne des études africaines* 50, no. 3 (2016): 415–35.
70. Ann H. Kelly et al., "'Like Sugar and Honey': Embedded Ethics of a Larval Control Project in The Gambia," *Social Science and Medicine* 70, no. 12 (2010): 1912–19.
71. Katherine A. Muldoon et al., "Supporting Southern-Led Research: Implications for North-South Research Partnerships," *Canadian Journal of Public Health* 103, no. 2 (March 2012): 128–31.
72. Rajan, "Experimental Values."
73. Johanna Crane, "Adverse Events and Placebo Effects: African Scientists, HIV, and Ethics in the 'Global Health Sciences,'" *Social Studies of Science* 40, no. 6 (2010): 846.
74. Johanna Crane, *Scrambling for Africa: AIDS, Expertise, and the Rise of American Global Health Science* (Ithaca, NY: Cornell University Press, 2013), 7. There is also a longer

history to the complicated mobilization of African difference and commensurability: see Kirsten Moore-Sheeley, "The Products of Experiment: Changing Conceptions of Difference in the History of Tuberculosis in East Africa, 1920s–1970s," *Social History of Medicine*, June 28, 2017, https://doi.org/10.1093/shm/hkx048.

75. Harry M. Marks, *The Progress of Experiment: Science and Therapeutic Reform in the United States, 1900–1990* (Cambridge: Cambridge University Press, 1997).
76. The subject/object split is fundamental to science, as explicated by feminist scholars of science such as Evelyn Fox Keller; see her *Reflections on Gender and Science* (New Haven, CT: Yale University Press, 1985).
77. This framing was one that I started thinking about as part of a conference panel, "Technoscience Circulation in the Global South: Beyond Diffusion and Translation," organized by Logan D. A. Williams, at the 2013 meeting of the Society for Social Studies of Science in San Diego, CA. For her further elaboration of this theme, see Logan D. A. Williams, *Eradicating Blindness: Global Health Innovation from South Asia* (London: Palgrave Macmillan, 2019).
78. Warwick Anderson, "Postcolonial Technoscience," *Social Studies of Science* 32, nos. 4–5 (October–December 2002): 644.
79. Ibid.

CHAPTER 2

1. The eucalyptus trees are themselves a fragrant relic of colonial history: they were brought from Australia because they grew quickly and, in addition to their aesthetic appeal, could provide both structural timber and solar protection for the mining industry and explosives manufacture. Rocco Brosman, *Modderfontein Village Development Heritage Impact Assessment, Annexure 1: The History Report* (Johannesburg: Rocco Brosman, February 2010), 58, accessed December 26, 2017, https://modderconserve.files.wordpress.com/2010/05/modderfontein-hia-annexure-1-historical-report.pdf.
2. Lucille Davie, "Peaceful Park Hides Explosive Past," 2005, accessed August 9, 2010, http://www.joburg.org.za/content/view/1109/168.
3. In its current iteration, the reserve is supported by a volunteer organization called the Modderfontein Conservation Society (https://modderconserve.wordpress.com/, accessed December 26, 2017) and managed by a nonprofit called the Endangered Wildlife Trust; see *Conservation Matters* 5 (August–September 2017): 30, accessed December 26, 2017, http://opus.sanbi.org/bitstream/123456789/5589/1/Magazine%20August%202017_final.pdf.
4. These gated communities are very typical of Johannesburg; see Charlotte Lemanski, Karina Landman, and Matthew Durington, "Divergent and Similar Experiences of 'Gating' in South Africa: Johannesburg, Durban and Cape Town," *Urban Forum* 19, no. 2 (2008): 133–58.
5. Relatedly, there was also a government lab doing drug discovery research there, CSIR Biosciences, with which iThemba would pool resources early on. CSIR Biosciences has since ceased drug discovery work and moved to Pretoria.
6. Indeed, one economic analyst offers as conventional wisdom: "given relatively high wages, low skills, distance from markets, etc., cheap electricity is probably the only competitive advantage that South Africa has." Seeraj Mohamed, "The Energy-Intensive Sector: Considering South Africa's Comparative Advantage in Cheap Energy," *Trade and Industrial Policy Strategies (TIPS) Forum*, 1998, 6–7, accessed

June 18, 2016, http://www.tips.org.za/research-archive/annual-forum-papers/1998/item/download/8_6b0779ade5b600f6ce8e9ff3c110e022.

7. Lack of reliable availability of electricity is a significant barrier to pharmaceutical manufacture in Nigeria, for example, as described by Kristin Peterson, *Speculative Markets: Drug Circuits and Derivative Life in Nigeria* (Durham, NC: Duke University Press, 2014), 123.
8. Marissa Mika, "Fifty Years of Creativity, Crisis, and Cancer in Uganda," *Canadian Journal of African Studies / Revue canadienne des études africaines* 50, no. 3 (2016): 397.
9. Ann Laura Stoler, "'The Rot Remains': From Ruins to Ruination," in *Imperial Debris: On Ruins and Ruination*, ed. Ann Laura Stoler (Durham, NC: Duke University Press, 2013), 2.
10. Karl Marx, *Capital*, vol. 3 (1894), 72, accessed October 2, 2016, https://www.marxists.org/archive/marx/works/download/pdf/Capital-Volume-III.pdf. For discussion, see Esther Leslie, *Nature, Art and the Chemical Industry* (London: Reaktion Books, 2005), 84.
11. Alan Dronsfield, "Pain Relief: From Coal Tar to Paracetamol," Royal Society for Chemistry: Education in Chemistry, July 2005, accessed July 2, 2016, http://www.rsc.org/education/eic/issues/2005July/painrelief.asp.
12. John J. Beer, "Coal Tar Dye Manufacture and the Origins of the Modern Industrial Research Laboratory," *Isis* 49, no. 2 (1958): 123–31. See also Peter J. T. Morris, *The Matter Factory: A History of the Chemistry Laboratory* (London: Reaktion Books, 2015), 246–47.
13. Judy Slinn, "The Development of the Pharmaceutical Industry," in *Making Medicines: A Brief History of Pharmacy and Pharmaceuticals*, ed. Stuart Anderson (London: Pharmaceutical Press, 2005), 162.
14. Vernard L. Foley and Patrick F. Belcastro, "William Brockedon and the Mechanization of Pill and Tablet Manufacture: From Bullets to Pills," *Pharmaceutical Technology*, September 1987, 110. See discussion in Emily Martin, "The Pharmaceutical Person," *BioSocieties* 1, no. 1 (2006): 275.
15. Foley and Belcastro, "William Brockedon," 112.
16. See E. Martin, "Pharmaceutical Person," 275.
17. The few exceptions are potassium chlorate, sodium bicarbonate, and potassium bicarbonate. William A. Jackson, "From Electuaries to Enteric Coating: A Brief History of Dosage Forms," in Anderson, *Making Medicines*, 212.
18. Nobel likely learned about nitroglycerin through a Russian tutor he shared with Sobrero. Nils Ringertz, "Alfred Nobel—His Life and Work," *Nature Reviews Molecular Cell Biology* 2, no. 12 (December 2001): 926. See also Linda Culp Holmes and Frederick J. DiCarlo, "Nitroglycerin: The Explosive Drug," *Journal of Chemical Education* 48, no. 9 (1971): 573–76.
19. Jaime Wisniak, "The Development of Dynamite: From Braconnot to Nobel," *Educación química* 19, no. 1 (2008): 79.
20. David Taylor, "The Pharmaceutical Industry and the Future of Drug Development," in *Pharmaceuticals in the Environment*, ed. R. E. Hester and R. M. Harrison (London: Royal Society of Chemistry, 2015), 5.
21. Jie Jack Li, *Laughing Gas, Viagra, and Lipitor: The Human Stories behind the Drugs We Use* (New York: Oxford University Press, 2006), 80.
22. Anthony Butler and Rosslyn Nicholson, *Life, Death and Nitric Oxide* (Cambridge: Royal Society of Chemistry, 2003), 32–33.
23. Timothy Scott-Burden, "Nitric Oxide Leads to Prized NObility: Background to the Work of Ferid Murad," *Texas Heart Journal* 26, no. 1 (1999): 1.

24. Stephen L. Archer, "The Making of a Physician-Scientist—the Process Has a Pattern: Lessons from the Lives of Nobel Laureates in Medicine and Physiology," *European Heart Journal* 28 (2007): 511.
25. Stephen R. Brown, *A Most Damnable Invention: Dynamite, Nitrates, and the Making of the Modern World* (Toronto: Penguin Group, 2005).
26. Ibid., 162.
27. See Pollock, *Medicating Race*, esp. 169–73; Persson, "Incorporating Pharmakon."
28. E. Martin, "Pharmaceutical Person," 282.
29. Li, *Laughing Gas, Viagra, and Lipitor*, 8–10.
30. Joanna Behrens, "The Dynamite Factory: An Industrial Landscape in Late-Nineteenth-Century South Africa," *Historical Archaeology* 39, no. 3 (2005): 61–74; Alan Patrick Cartwright, *The Dynamite Company: The Story of African Explosives and Chemical Industries* (Cape Town: Purnell and Sons, 1964).
31. Ben Fine and Zavareh Rustomjee, *The Political Economy of South Africa: From the Minerals-Energy Complex to Industrialization* (Boulder, CO: Westview Press, 1996), 5.
32. Comaroff and Comaroff, *Theory from the South*, 54.
33. Stuart Jones and André Müller, *The South African Economy, 1910–1990* (New York: St. Martin's Press, 1992), 12.
34. Ibid., 13.
35. Clive Mennell, quoted in Fine and Rustomjee, *Political Economy of South Africa*, 141.
36. Ibid.
37. For a discussion of the ways that this operated in the German dyestuffs industry, which diversified into pharmaceuticals quite early, see Jonathan Liebenau, "Ethical Business: The Formation of the Pharmaceutical Industry in Britain, Germany and the United States before 1914," *Business History* 30 (1988): 116–29, esp. 118.
38. R. E. Altona, "The Disposal of Industrial Effluents on Pastures," *Proceedings of the Annual Congresses of the Grassland Society of Southern Africa* 2, no. 1 (1967): 143–46.
39. Cartwright, *Dynamite Company*.
40. Struggles over guns provide their own fascinating route into the history of colonialism and exploitation in South Africa. See William Kelleher Storey, *Guns, Race, and Power in Colonial South Africa* (New York: Cambridge University Press, 2008).
41. Nancy L. Clark, *Manufacturing Apartheid: State Corporations in South Africa* (New Haven, CT: Yale University Press, 1994), 20–21; Cartwright, *Dynamite Company*, 46; Jones and Müller, *South African Economy*, 67–68. "In 1893, the government perceived an opportunity to benefit from the importation, production and sale of explosives and made this a state monopoly. In the following year the Nobel Dynamite Trust was given the licence to operate the monopoly, which led to the construction of an explosives factory at Modderfontein, near Johannesburg. The factory, which came into production in 1896, was the largest commercial explosives factory in the world. Its production of nitric acid, sulphuric acid and other ingredients for the manufacture of explosives heralded the birth of the chemical industry in South Africa" (Jones and Müller, *South African Economy*, 68). See also Stanley Trapido, "Imperialism, Settler Identities, and Colonial Capitalism," in *The Cambridge History of South Africa*, vol. 2, *1885–1994*, ed. Robert Ross, Anne Kelk Mager, and Bill Nasson (Cambridge: Cambridge University Press, 2011), 66–101, esp. 83.
42. Clark, *Manufacturing Apartheid*, 20–21.
43. Ibid., 21.
44. Cartwright, *Dynamite Company*, 46.
45. Ibid., 43.

46. African Explosives and Chemical Industries, "History," accessed January 18, 2016, http://www.aeci.co.za/aa_history.php.
47. The scope and legacy of the Rhodes Must Fall campaign are highly contested. See Siona O'Connell, "A Search for the Human in the Shadow of Rhodes," *Ufahamu: A Journal of African Studies* 38, no. 3 (2015): 11–14.
48. William Kelleher Storey, "Cecil Rhodes and the Making of a Sociotechnical Imaginary for South Africa," in Jasanoff and Kim, *Dreamscapes of Modernity*, 34.
49. Ibid., 47.
50. Edwards and Hecht, "History and the Technopolitics of Identity," 622.
51. V. Y. Mudimbe, *The Invention of Africa: Gnosis, Philosophy, and the Order of Knowledge* (Bloomington: Indiana University Press, 1988), 79.
52. This seems to be the case in an ad for a local Indian generic-pharmaceutical company's cardiovascular medicine that features a figuration of a molecule that imitates the style used by Sasol for such illustrations. See a reproduction of the ad and discussion in Kottakkunnummal, "Social Life of Indian Generic Pharmaceuticals in Johannesburg," 31.
53. Second after only Portugal. Donald G. McNeil Jr., "South Africa's Bitter Pill for World's Drug Makers," *New York Times*, March 29, 1998, http://www.nytimes.com/1998/03/29/business/south-africa-s-bitter-pill-for-world-s-drug-makers.html.
54. William W. Fisher III and Cyrill P. Rigamonti, "The South Africa AIDS Controversy: A Case Study in Patent Law and Policy," Working Paper, Harvard Law School, February 10, 2005, accessed January 31, 2016, http://cyber.law.harvard.edu/people/tfisher/South%20Africa.pdf.
55. Linda-Gail Bekker et al., "Provision of Antiretroviral Therapy in South Africa: The Nuts and Bolts," *Antiviral Therapy* 19, no. S3 (2014): 109.
56. The idea of noninnocence is developed by feminist science and technology studies theorist Donna Haraway in "A Cyborg Manifesto: Science, Technology, and Socialist-Feminism in the Late Twentieth Century," in *Simians, Cyborgs, and Women: The Reinvention of Nature* (New York: Routledge, 1991), 149–81. For Haraway, the cyborg is the bastard child of militarism and capitalism, but it can be a liberatory figure because it is not necessarily faithful to its origins.
57. David M. Matsinhe, *Apartheid Vertigo: The Rise in Discrimination against Africans in South Africa* (Farnham, UK: Ashgate, 2011), 27.
58. Antina von Schnitzler, "Traveling Technologies: Infrastructure, Ethical Regimes, and the Materiality of Politics in South Africa," *Cultural Anthropology* 28, no. 4 (2013): 670–93.
59. Nancy L. Clark, "Structured Inequality: Historical Realities of the Post-apartheid Economy," *Ufahamu: A Journal of African Studies* 38, no. 1 (2014): 97.
60. Crispen Chinguno, "Marikana Massacre and Strike Violence Post-apartheid," *Global Labour Journal* 4, no. 2 (2013): 160–66.
61. Terry Bell, "The Marikana Massacre: Why Heads Must Roll," *New Solutions: A Journal of Environmental and Occupational Health Policy* 25, no. 4 (2016): 440.
62. Keith Breckenridge, "Marikana and the Limits of Biopolitics: Themes in the Recent Scholarship of South African Mining," *Africa* 84, no. 1 (February 2014): 151.
63. James Ferguson, *Expectations of Modernity: Myths and Meanings of Urban Life on the Zambian Copperbelt* (Berkeley: University of California Press, 1999).
64. The drug discovery and development company that was founded shortly after iThemba and outlasted it, H3D, was directed by Zambian Kelly Chibale.
65. Kim Fortun, *Advocacy after Bhopal: Environmentalism, Disaster, New Global Orders*

(Chicago: University of Chicago Press, 2001); Deboleena Roy, "Germline Ruptures: Methyl Isocyanate Gas and the Transpositions of Life, Death, and Matter in Bhopal," paper presented at University of California, Los Angeles, Center for the Study of Women, Life(Un)Ltd. Series, November 2013.

66. Michelle Murphy, "Distributed Reproduction, Chemical Violence, and Latency," *Scholar and Feminist Online* 11, no. 3 (Summer 2013)
67. Noémi Tousignant, *Edges of Exposure: Toxicology and the Problem of Capacity in Postcolonial Senegal* (Durham, NC: Duke University Press, 2018).

CHAPTER 3

1. Mike Phflanz, "Nelson Mandela: The Quiet Street Where He Died Is Filled with Clapping and Singing," *Telegraph* (London), December 6, 2013, accessed October 21, 2017, http://www.telegraph.co.uk/news/worldnews/nelson-mandela/10501578/Nelson-Mandela-the-quiet-street-where-he-died-is-filled-with-clapping-and-singing.html.
2. Nelson Mandela, "Forward," in *Building a New South Africa*, vol. 3, *Science and Technology Policy: A Report from the Mission on Science and Technology Policy for a Democratic South Africa*, ed. Marc Van Ameringen (Ottowa, Canada: International Development Research Centre, 1995), vii–viii.
3. The popular enthusiasm among the South African crowds in response to the presence of international leaders, especially Barack Obama, was generally missed in international media coverage, which treated their presence as a spectacle. See Katie M. Milner and Nancy Baym, "The Selfie of the Year of the Selfie: Reflections on a Media Scandal," *International Journal of Communication* 9 (2015): 1701–15. That said, the funeral was certainly a media event for South Africans as well. See Martha Evans, "The Last TV Star? Nelson Mandela's Funeral Broadcast, Social Media, and the Future of Media Events," in *Global Perspectives on Media Events in Contemporary Society*, ed. Andrew Fox (Hershey, PA: Information Science Reference, 2016), 141–57.
4. See William Gumede's thoughtful comments on Mandela's democratic morality in his introduction to a 2013 reprinting of Nelson Mandela's *No Easy Walk to Freedom* (Paarl, RSA: Kwela Books, 2013), 7–28. On the postpresidency period, see John Daniel, "Soldiering On: The Post-presidential Years of Nelson Mandela," in *Legacies of Power: Leadership Change and Former Presidents in African Politics*, ed. Henning Melber and Roger Southall (Cape Town: HSRC Press, 2006), 26–50.
5. Social science literature has characterized the 2010 feeling as "pride/euphoria," which I would suggest is fertile ground for hope. Heather J. Gibson et al., "Psychic Income and Social Capital among Host Nation Residents: A Pre–Post Analysis of the 2010 FIFA World Cup in South Africa," *Tourism Management* 44 (2014): 113–22. The meaning of the fleeting nationalism and pan-Africanism engendered by the 2010 FIFA World Cup has been highly contested: Sabelo J. Ndlovu-Gatsheni, "Pan-Africanism and the 2010 FIFA World Cup in South Africa," *Development Southern Africa* 28, no. 3 (September 2011): 401–13.
6. Christopher Saunders, "Perspective on the Transition from Apartheid to Democracy in South Africa," *South African Historical Journal* 51, no. 1 (2004): 159–66.
7. Mandisa Mbali, *South African AIDS Activism and Global Health Politics* (London: Palgrave Macmillan, 2013).
8. Cooper, "On Pharmaceutical Empire," 52–53.
9. Médecins sans frontières Access Campaign, "1998: Big Pharma versus Nelson Man-

dela," January 2009, accessed February 3, 2016, http://www.msfaccess.org/content/1998-big-pharma-versus-nelson-mandela.

10. W. F. J. Steenkamp, "The Pharmaceutical Industry in South Africa," *South African Journal of Economics* 47, no. 1 (1979): 46–57.
11. Fisher and Rigamonti, "South Africa AIDS Controversy," 2–3.
12. World Trade Organization, "Fiftieth Anniversary of the Multilateral Trading System," https://www.wto.org/english/thewto_e/minist_e/min96_e/chrono.htm.
13. See Emilie Cloatre, *Pills for the Poorest: An Exploration of TRIPS and Access to Medication in Sub-Saharan Africa* (London: Palgrave Macmillan, 2013), 5.
14. India continues to make the vast majority of HIV drugs consumed in South Africa, many of which were generic worldwide by the time that India had to comply with TRIPS.
15. Fisher and Rigamonti, "South Africa AIDS Controversy," 6.
16. Patrick Bond, "Globalization, Pharmaceutical Pricing, and South African Health Policy: Managing Confrontation with US Firms and Politicians," *International Journal of Health Services* 29, no. 4 (October 1999): 765.
17. David Walwyn, personal e-mail communication, February 11, 2016.
18. Quoted in Fisher and Rigamonti, "South Africa AIDS Controversy," 9.
19. Helen Schneider, "On the Fault-Line: The Politics of AIDS Policy in Contemporary South Africa," *African Studies* 61, no. 1 (2002): 145–67.
20. Mbali, *South African AIDS Activism and Global Health Politics*, 2.
21. Deborah Posel makes an analogous point with regard to sexuality: "*if* AIDS was sexually transmitted, then—given the scale of the problem—the consequences would be intolerable." Deborah Posel, "Sex, Death, and the Fate of the Nation: Reflections on the Politicization of Sexuality in Post-apartheid South Africa," *Africa: Journal of the International African Institute* 75, no. 2 (2005): 125–53.
22. Cooper, "On Pharmaceutical Empire"; Fassin, *When Bodies Remember*.
23. Mbali, *South African AIDS Activism and Global Health Politics*; see, e.g., 11.
24. Jasanoff and Kim, *Dreamscapes of Modernity*.
25. See Gillespie, "Reclaiming Nonracialism," 65.
26. For a broadly pessimistic interpretation of the investment in nonracialism in South Africa and elsewhere, see David Theo Goldberg, *The Threat of Race: Reflections on Racial Neoliberalism* (Malden, MA: Wiley Blackwell, 2009).
27. Sarah Nuttall, "Subjectivities of Whiteness," *African Studies Review* 44, no. 2 (2001): 118.
28. The notion of using exchange programs as a way to forge a community of South African students who would be able to build a post-apartheid society is by no means Fisher's alone. For example, it was also part of the mission of the South African Student Exchange Program that brought 1,500 South Africans to the United States for study between 1979 and 1995—students who would play a role in anti-apartheid activism in both the United States and South Africa and would become leaders in post-apartheid South Africa. Thomas McClendon and Pamela Scully, "The South African Student Exchange Program: Anti-apartheid Activism in the Era of Constructive Engagement," *Safundi: The Journal of South African and American Studies* 16, no. 1 (2015): 1–27.
29. I interviewed Julian Walsh via Skype on May 6, 2013.
30. W. E. B. Du Bois, *The Souls of Black Folk* (Chicago: A. C. McClurg, 1903).
31. Cooper, "On Pharmaceutical Empire."

32. For discussion of this sense of Africanization, see, e.g., Stanley Shaloff, "The Africanization Controversy in the Gold Coast, 1926–1946," *African Studies Review* 17, no. 3 (December 1974): 493–504.
33. Cherry, "South African Science," 728.
34. Lyn Schumaker, *Africanizing Anthropology: Fieldwork, Networks, and the Making of Cultural Knowledge in Central Africa* (Durham, NC: Duke University Press, 2001).
35. Esperanza Brizuela-Garcia, "The History of Africanization and the Africanization of History," *History in Africa* 33 (2006): 85–100.
36. One might argue that this mission is continuous with colonial imperatives to control the population rather than postcolonial imperatives to serve the population. That would put efforts to discover new drugs for TB, HIV, and malaria into the long trajectory of colonial medicine in Africa, as described by, for example, Megan Vaughan, *Curing Their Ills: Colonial Power and African Illness* (Redwood City, CA: Stanford University Press, 1991). However, that frame would not resonate with any of the actors involved with iThemba. This may be in part because the South African state and industry have historically responded to black disease predominantly through segregation, as explored in Randall M. Packard, *White Plague, Black Labor: Tuberculosis and the Political Economy of Health and Disease in South Africa* (Berkeley: University of California Press, 1989).
37. The history of the local and global pharmaceutical industries in South Africa under sanctions is a complicated one. See S. Prakash Sethi and Oliver F. Williams, "Eli Lilly and Company," in *Economic Imperatives and Ethical Values in Global Business: The South African Experience and International Codes Today* (Boston: Kluwer Academic, 2000), 127–38; Britt Akermann and Faiz Kermani, "The Development of the South African Biotech Sector," *Journal of Commercial Biotech Sector* 12, no. 2 (2006): 111–19.
38. C. te W. Naudé and J. M. Luis, "An Industry Analysis of Pharmaceutical Production in South Africa," *South African Journal of Business Management* 44, no. 1 (2013): 33–46.
39. Radhamany Sooryamoorthy, "Science and Scientific Collaboration in South Africa: Apartheid and After," *Scientometrics* 84 (2010): 373–90.
40. Susan Craddock, "Drug Partnerships and Global Practices," *Health and Place* 18, no. 3 (2012): 481–89.
41. World Bank, *World Development Report 2013: Jobs* (Washington, DC: World Bank, 2012), https://doi.org/10.1596/978-0-8213-9575-2.
42. Michael M. J. Fischer, "Lively Capital and Translational Research," in Rajan, *Lively Capital*, 419; David Zweig, Chung Siu Fung, and Donglin Han, "Redefining the 'Brain Drain': China's Diaspora Option," *Science, Technology and Society* 13, no. 1 (2008): 1–33.
43. Itty Abraham, "The Contradictory Spaces of Postcolonial Techno-science," *Economic and Political Weekly* 41, no. 3 (2006): 217.
44. Leon de Kock, "South Africa in the Global Imaginary: An Introduction," *Poetics Today* 22, no. 2 (2001): 289.
45. Mandela, "Forward."
46. Ibid.
47. "South Africa's Opportunity," *Nature* 463 (February 2010): 709; Salim S. Abdool Karim and Quarraisha Abdool Karim, "AIDS Research Must Link to Local Policy," *Nature* 463 (February 2010): 734.

CHAPTER 4

1. In a sense, my approach resonates with the methodology of intersectionality that feminist sociologist Leslie McCall has termed *intracategorial complexity*, even though protecting the identity of individual scientists precludes me from exploring gender itself as a category of intersectional analysis within the category "African scientist." Leslie McCall, "The Complexity of Intersectionality," *Signs: Journal of Women in Culture and Society* 30, no. 3 (2005): 1771–1800.
2. Damien Droney, "Ironies of Laboratory Work during Ghana's Second Age of Optimism," *Cultural Anthropology* 29, no. 2 (2014): 369.
3. Nolwazi Mkhwanazi, "Medical Anthropology in Africa: The Trouble with a Single Story," *Medical Anthropology* 35, no. 2 (2016): 193–202. She is drawing on the evocative phrasing of Nigerian novelist Chimamanda Ngozi Adichie, "The Danger of a Single Story," TED Talk, 2009, accessed July 29, 2016, https://www.ted.com/talks/chimamanda_adichie_the_danger_of_a_single_story?language=en.
4. Donna Haraway, "Situated Knowledges: The Science Question in Feminism and the Privilege of Partial Perspective," *Feminist Studies* 14, no. 3 (1988): 584.
5. Hilary Rose, "Hand, Brain, and Heart: A Feminist Epistemology for the Natural Sciences," *Signs* 9, no. 1 (Autumn 1983): 73–90.
6. Maria Puig de la Bellacasa, "Matters of Care in Technoscience: Assembling Neglected Things," *Social Studies of Science* 41, no. 1 (2011): 85–106.
7. See Susan Cozzens, "Distributive Justice in Science and Technology Policy," *Science and Public Policy* 34, no. 2 (March 2007): 88–89. "The actual distribution of benefits under a purely utilitarian S&T policy approach would be out of the hands of those producing knowledge, and in the hands of social policy, the market, or philanthropy, depending on the society where the policy was operating" (93).
8. UNAIDS, "GAP Report 2014," 17, http://www.unaids.org/sites/default/files/media_asset/UNAIDS_Gap_report_en.pdf.
9. UNAIDS, "South Africa HIV and AIDS Estimates (2014)," accessed September 17, 2015, http://www.unaids.org/en/regionscountries/countries/southafrica.
10. Statistics South Africa, "Mortality and Causes of Death in South Africa, 2013: Findings from Death Notification," 27, accessed September 17, 2015, http://www.statssa.gov.za/publications/P03093/P030932013.pdf.
11. Steven Robins, "From 'Rights' to 'Ritual': AIDS Activism in South Africa," *American Anthropologist* 108, no. 2 (June 2006): 321.
12. Robins, "'Long Live Zackie, Long Live,'" 663.
13. On the problems of individualizing responsibility for HIV treatment, see Kiran Pienaar, "Claiming Rights, Making Citizens: HIV and the Performativity of Biological Citizenship," *Social Theory and Health* 14, no. 2 (May 2016): 149–68.
14. Kevin O'Brien, "Drug Companies and AIDS in Africa," *America Magazine*, November 25, 2002, accessed November 10, 2017, https://www.americamagazine.org/issue/413/article/drug-companies-and-aids-africa.
15. Annamarie Bindenagel Šehović, *HIV/AIDS and the South African State: Sovereignty and the Responsibility to Respond* (Farnham, UK: Ashgate, 2014).
16. Alondra Nelson, "The Inclusion-and-Difference Paradox," review of *Inclusion: The Politics of Difference in Medical Research*, by Steven Epstein, *Social Identities* 15 (2009): 742–43.
17. Wen Hua Kuo, "Understanding Race at the Frontier of Pharmaceutical Regulation: An Analysis of the Racial Difference Debate at the ICH," *Journal of Law, Medicine, and Ethics* 36, no. 3 (September 2008): 498–505.

18. See, e.g., Laura A. Foster, "A Postapartheid Genome: Genetic Ancestry Testing and Belonging in South Africa," *Science, Technology, and Human Values* 41, no. 6 (2016): 1015–36.
19. That is, the belonging is not a molecular biopolitics. Cf. Nikolas Rose, *The Politics of Life Itself: Biomedicine, Power, and Subjectivity in the Twenty-First Century* (Princeton, NJ: Princeton University Press, 2006).
20. In this sense, the perspective of this scientist and colleagues in South Africa is well aligned with science and technology studies broadly and decolonial science and technology studies in particular. As Kim TallBear (*Native American DNA*, 203) argues: "Whether we attempt to disown recognition of these politics as an imposition *onto* science or whether we strive to understand how these politics always already condition our scientific practices will make an important difference in whether the genome sciences evolve in more or less democratic ways throughout the twenty-first century."
21. Gayatri Chakravorty Spivak, "Can the Subaltern Speak?" (abbreviated by the author), in *The Post-colonial Studies Reader*, ed. Bill Ashcroft, Gareth Griffiths, and Helen Tiffin (New York: Routledge, 1995), 28.
22. Michel Callon, "Some Elements of a Sociology of Translation: Domestication of the Scallops and the Fishermen of St. Brieuc Bay," in *Power, Action and Belief: A New Sociology of Knowledge*, ed. John Law (London: Routledge and Kegan Paul, 1986), 196–233.
23. Hayden, *When Nature Goes Public*.
24. Lack of access to food is a major "unknown known" of health research in Africa, as Geissler ("Public Secrets in Public Health," 19) points out.
25. Sandra Harding, "Beyond Postcolonial Theory: Two Undertheorized Perspectives in Science and Technology," introduction to Harding, *Postcolonial Science and Technology Studies Reader*, 14–15.
26. See Pollock, *Medicating Race*.
27. The implicit opposition between the pharmaceutical and the social is not completely absent from global health literature of pharmaceuticalization, such as Biehl, "Pharmaceuticalization," but is far more fundamental and explicit in pharmaceuticalization critiques focused on rich countries. See Simon J. Williams, Paul Martin, and Jonathan Gabe, "The Pharmaceuticalisation of Society? A Framework for Analysis," *Sociology of Health and Illness* 33, no. 5 (2011): 710–25; Anne Pollock and David S. Jones, "Coronary Artery Disease and the Contours of Pharmaceuticalization," *Social Science and Medicine* 131 (April 2015): 221–27. For discussion of the gaps between pharmaceuticalization as framed by sociologists analyzing the Global North and by anthropologists analyzing the Global South, see Susan E. Bell and Anne E. Figert, "Medicalization and Pharmaceuticalization at the Intersections: Looking Backward, Sideways and Forward," *Social Science and Medicine* 75 (2012): 2131–33.
28. Sandra Harding, "Postcolonial and Feminist Philosophies of Science and Technology: Convergences and Dissonances," *Postcolonial Studies* 12, no. 4 (2009): 403.
29. Questions of whether ongoing and in some ways increasing segregation constitutes a "new apartheid" are longstanding—see, e.g., Keith S. O. Beavon, "Northern Johannesburg: Part of the 'Rainbow' or Neo-apartheid City in the Making?," *Mots pluriels* 13 (2000), accessed June 25, 2016, http://motspluriels.arts.uwa.edu.au/MP1300kb.html; Owen Crankshaw, "Race, Space, and the Post-Fordist Spatial Order of Johannesburg," *Urban Studies* 45, no. 8 (July 2008): 1692–711.
30. It can be easy to forget how important proximity to family and friends can be for

scientists everywhere: both iThemba cofounder Dennis Liotta and his collaborator in the discovery of transformative drugs for HIV and hepatitis C Ray Schinazi initially took jobs at Emory in Atlanta because they had family members in the city. Liotta told me in an oral history in 2012 about the appeal of living near his older brother who was on the faculty at Georgia Tech, and Schinazi followed his uncle to Emory. Jon Cohen, "King of the Pills: Raymond Schinazi's Handful of Lifesaving Drugs Has Earned Him Riches, Esteem, and a Dose of Enmity," *Science* 348, no. 6235 (2015): 624.

31. Haraway, "Situated Knowledges," 586.
32. Ibid., 587.
33. Anderson, "From Subjugated Knowledge to Conjugated Subjects," 389.
34. Homi Bhabha, "Of Mimicry and Man: The Ambivalence of Colonial Discourse," in *The Location of Culture* (London: Routledge, 1994), 85–92.
35. "Freedom Charter: Congress of the People," in *The South Africa Reader: History, Culture, Politics*, ed. Clifton Crais and Thomas V. McClendon (Durham, NC: Duke University Press, 2013), 320.
36. Saul Dubow, "South Africa and South Africans: Nationality, Belonging, Citizenship," in Ross, Mager, and Nasson, *Cambridge History of South Africa*, 2:63.
37. Preamble to the Constitution of the Republic of South Africa, 1996, 1, accessed November 4, 2017, http://www.justice.gov.za/legislation/constitution/SAConstitution-web-eng.pdf.
38. Jean Comaroff and John L. Comaroff, "Law and Disorder in the Postcolony: An Introduction," in *Law and Disorder in the Postcolony*, ed. Jean Comaroff and John L. Comaroff (Chicago: University of Chicago Press, 2006), 24.
39. "Non-racialism" is the language of the Constitution, which declares that South Africa is a sovereign, democratic state founded on the values of human dignity, "non-racialism and non-sexism," rule of law, and universal suffrage (Constitution of the Republic of South Africa, 3).
40. Dubow, "South Africa and South Africans," 64.
41. Thabo Mbeki, "I Am an African," in Crais and McClendon, *South Africa Reader*, 475–80.
42. Foster, "Postapartheid Genome," 1022.
43. Nelson Mandela, "Inaugural Address," in Crais and McClendon, *South Africa Reader*, 471. The landscape is prominent in twentieth-century South African political discourse; see Jennifer Beningfield, *The Freighted Land: Land, Landscape, and Politics in South Africa in the Twentieth Century* (London: Routledge, 2006).
44. Mandela, "Inaugural Address," 472.
45. Ibid. The use of "we" rather than "I" is very characteristic of Mandela's speech and writing. See Rob Nixon, *Slow Violence and the Environmentalism of the Poor* (Cambridge, MA: Harvard University Press, 2011), 143.
46. Matsinhe, *Apartheid Vertigo*, 61.
47. Tilley, *Africa as a Living Laboratory*.
48. Kwame Anthony Appiah, *Cosmopolitanism: Ethics in a World of Strangers* (New York: Norton, 2010); see, for example, his comments on his father's final message (xviii).
49. Cindy Patton, "Inventing 'African AIDS,'" *new formations* 10 (Spring 1990): esp. 30.
50. Binyavanga Wainaina, "How to Write about Africa," *Granta* 92 (2006), https://granta.com/how-to-write-about-africa/.
51. Edward Said, *Orientalism* (New York: Pantheon Books, 1978), 3.
52. See Kelly and Geissler, *Value of Transnational Medical Research*.

53. Ferguson, *Global Shadows*, 5.
54. Mudimbe, *Invention of Africa*, x.
55. Clapperton Chakanetsa Mavhunga, *What Do Science, Technology, and Innovation Mean from Africa?* (Cambridge, MA: MIT Press, 2017), ix.
56. Ibid., 9–10.
57. Gabrielle Hecht, *Being Nuclear: Africans and the Global Uranium Trade* (Cambridge: Cambridge University Press, 2012), 21–22.
58. Mbembe, *On the Postcolony*, 242.
59. Jasanoff and Kim, "Containing the Atom," 123.
60. For more on this framing of beyond diffusion and translation, see L. Williams, *Circulating Sciences from Below*.

CHAPTER 5

1. Their access to these databases was itself part of their privilege relative to many scientists and others across the African continent. For discussion of these issues, including South Africa's betwixt-and-between status with regard to the digital, see Peter Limb, "The Digitization of Africa," *Africa Today* 52, no. 2 (Winter 2005): 3–19.
2. Mbembe, *On the Postcolony*, 242.
3. This point is foundational to the anthropology of pharmaceuticals, as canonically explored in Van der Geest, Whyte, and Hardon, "Anthropology of Pharmaceuticals."
4. The term "material-semiotic" comes from Donna Haraway, who uses the term in many places, including "Situated Knowledges," 588.
5. Cori Hayden, "Rethinking Reductionism; or, The Transformative Work of Making the Same," *Anthropological Forum* 22, no. 3 (November 2012): 274.
6. P. Morris, *Matter Factory*, esp. 97.
7. Evelyn Fox Keller, *A Feeling for the Organism: The Life and Work of Barbara McClintock* (San Francisco: W. H. Freeman, 1983), 198.
8. The study of primatologists has been foundational in posthuman engagements with science, ever since Donna Haraway's landmark *Primate Visions: Gender, Race, and Nature in the World of Modern Science* (New York: Routledge, 1990). Entomology is among the most developed topics of the more-than-human in science in Africa; see Uli Biesel, Ann H. Kelly, and Noémi Tousignant, "Knowing Insects: Hosts, Vectors, and the Companions of Science," *Science as Culture* 22, no. 1 (2013): 1–13.
9. Bernadette Bensaude-Vincent and Isabelle Stengers, *A History of Chemistry*, trans. Deborah van Dam (Cambridge, MA: Harvard University Press, 1996), 10.
10. Bernadette Bensaude-Vincent, "Philosophy of Chemistry or Philosophy with Chemistry?," *HYLE: International Journal for the Philosophy of Chemistry* 20 (2014): 60.
11. Ibid., 69.
12. Haraway, "Cyborg Manifesto." For examples of such naturecultural investigations, see Banu Subramaniam, *Ghost Stories for Darwin: The Science of Variation and the Politics of Diversity* (Urbana: University of Illinois Press, 2014); Foster, *Reinventing Hoodia*.
13. Bensaud-Vincent, "Philosophy of Chemistry," 70.
14. Ibid.
15. Ibid., 71.
16. See Rosengarten, *HIV Interventions*.
17. Andrew Barry, "Pharmaceutical Matters: The Invention of Informed Materials," *Theory, Culture and Society* 22, no. 1 (2005): 52. Barry also points out, "Pharmaceutical companies do not just sell information, nor do they just sell material objects (drug

molecules). The molecules produced by pharmaceutical companies are more or less purified, but they are also enhanced and enriched through laboratory practice. The molecules produced by a pharmaceutical company are already part of a rich informational material environment, even before they are consumed" (59).

18. Ibid., 64.
19. Bensaude-Vincent and Stengers, *History of Chemistry*, 10.
20. Cherry, "South African Science," 727.
21. Bernadette Bensaud-Vincent and Jonathan Simon, *Chemistry: The Impure Science* (London: Imperial College Press, 2008), 5.
22. Bensaude-Vincent and Stengers, *History of Chemistry*, 5.
23. Chambers and Gillespie, "Locality in the History of Science," 231.
24. Annemarie Mol and John Law, "Regions, Networks and Fluids: Anaemia and Social Topology," *Social Studies of Science* 24, no. 4 (November 1994): 649.
25. Okeke, *Divining without Seeds*, 157.
26. Chambers and Gillespie, "Locality in the History of Science," 223.
27. For how this quality of South Africa becomes framed as a value-added property for biotech research partnerships, see Sara Al-Bader et al., "Small but Tenacious: South Africa's Biotech Sector," *Nature Biotechnology* 27, no. 5 (2009): 435.
28. Laura Foster (*Reinventing Hoodia*, esp. 82–84) has insightfully described this phenomenon with regard to the ways that South African scientists patented the isolated active compound of the hoodia plant, even as the hope invested in that molecule was never fully realized.
29. See chapter 2 of this volume, "In the Shadows of the Dynamite Factory."
30. Comaroff and Comaroff, *Theory from the South*, 1.
31. Walter W. Powell and Kaisa Snellman, "The Knowledge Economy," *Annual Review of Sociology* 30 (2004): 199.
32. Ibid., 200.
33. Indeed, angel investors in South Africa have been much less present in biotech spaces than the government has been. See Al-Bader et al., "Small but Tenacious," 441.
34. Nicholas Negroponte, *Being Digital* (New York: Knopf, 1995). See also Nicholas Negroponte, "Bits and Atoms," January 1, 1995, accessed June 30, 2016, http://web.media.mit.edu/~nicholas/Wired/WIRED3-01.html.
35. Negroponte, *Being Digital*, 12–13.
36. Negroponte, "Bits and Atoms."
37. Issues of labor and infrastructure have become more prominent recently in studies of computing, as, for example, in Lisa Nakamura, "Indigenous Circuits: Navajo Women and the Racialization of Early Electronic Manufacture," *American Quarterly* 66, no. 4 (December 2014): 919–41. However, it's worth noting that those in critical race studies of the digital have long foregrounded labor and infrastructure together with representation. A good example is Alondra Nelson and Thuy Linh N. Tu's edited collection *TechniColor: Race, Technology and Everyday Life* (New York: New York University Press, 2001).
38. Louise M. Bezuidenhout et al., "Beyond the Digital Divide: Towards a Situated Approach to Open Data," *Science and Public Policy* 44, no. 4 (2017): 473.
39. As of this writing, there were 131,841,062 valid US passports in circulation, well under half of those eligible for them. See US Department of State, Bureau of Consular Affairs, "Passport Statistics," accessed November 29, 2016, https://travel.state.gov/content/passports/en/passports/statistics.html.

40. US Embassy and Consulates in South Africa, "How to Apply," accessed November 29, 2016, https://za.usembassy.gov/visas/tourism-visitor/how-to-apply/.
41. That is, this drug discovery effort provides a counterpoint to the kind of hype around globalization epitomized by Thomas Friedman's *The World Is Flat: A Brief History of the Twenty-First Century* (New York: Farrar, Straus, and Giroux, 2005).
42. The "thinginess" here draws on the notion from the anthropology of pharmaceuticals—canonically in Van der Geest, Reynolds-White, and Hardon, "Anthropology of Pharmaceuticals," 154—but also has resonances with new materialisms, as in Jane Bennett's *Vibrant Matter: A Political Ecology of Things* (Durham, NC: Duke University Press, 2010).
43. Alain Pottage, "Law Machines: Scale Models, Forensic Materiality and the Making of Modern Patent Law," *Social Studies of Science* 41, no. 5 (2011): 634.
44. As Emilie Cloatre and Robert Dingwall argue, in some senses regulation is embedded in pills themselves. See their "'Embedded Regulation': The Migration of Objects, Scripts, and Governance," *Regulation and Governance* 7 (2013): 365–86.
45. On the profoundly material history of patents, see Pottage, "Law Machines."
46. See Pollock, *Medicating Race*, esp. chaps. 5 and 6, about generic thiazide-type diuretics and the branded combination of generic ingredients BiDil.
47. On the life cycle of pharmaceuticals, see Van der Geest, Whyte, and Hardon, "Anthropology of Pharmaceuticals." On the traffic between information and flesh, see Rosengarten, *HIV Interventions*.
48. Andrew Pollack, "Drug Goes from $13.50 a Tablet to $750, Overnight," *New York Times*, September 20, 2015, https://www.nytimes.com/2015/09/21/business/a-huge-overnight-increase-in-a-drugs-price-raises-protests.html.
49. McNeil, "South Africa's Bitter Pill for World's Drug Makers."
50. For example, even for generic drugs, in the United States each new manufacturer needs to be FDA approved. See historian of medicine Jeremy A. Greene's popular account of barriers to and possibilities for reform: "Can the Government Stop the Next Martin Shkreli?," *Slate*, March 22, 2016, accessed March 30, 2016, http://www.slate.com/articles/business/moneybox/2016/03/the_fda_wants_to_stop_the_next_martin_shkreli_by_speeding_up_the_approval.html.

CHAPTER 6

1. Amade M'charek points out, "Although the actor network approach has taught us about objects as spatial enactments, little attention has been paid to the temporality of objects and, more specifically, to how objects enact time." Amade M'charek, "Race, Time and Folded Objects: The HeLa Error," *Theory Culture Society* 31, no. 6 (2014): 32. She is building on the topological understanding of time explored by Michel Serres, with Bruno Latour, in *Conversations on Science, Culture, and Time* (Ann Arbor: University of Michigan Press, 1995).
2. Indian manufacturers produce 85 percent of the generic drugs procured in sub-Saharan Africa: Colleen V. Chien, "HIV/AIDS Drugs for Sub-Saharan Africa: How Do Brand and Generic Supply Compare?," *PLoS ONE* 2, no. 3 (2007): e278, https://doi.org/10.1371/journal.pone.0000278.
3. David J. Ripin et al., "Antiretroviral Procurement and Supply Chain Management," *Antiviral Therapy* 19, no. S3 (2014): 79–89.
4. Brand South Africa, "South Africa Pays Less for ARV Drugs," December 15, 2010, accessed December 23, 2017, https://www.brandsouthafrica.com/south-africa-fast-facts/health-facts/arvdrugs-151210.

5. Sekhejane and Pelletan, "HIV and AIDS Triumph and Struggles," 98.
6. Bada Pharasi and Jacqui Miot, "Medicines Selection and Procurement in South Africa," *South African Health Review*, 2013, 117–85.
7. For discussion of India's path along these lines, see, e.g., Dinar Kale and Steve Little, "From Imitation to Innovation: The Evolution of R&D Capabilities and Learning Processes in the Indian Pharmaceutical Industry," *Technology Analysis and Strategic Management* 19, no. 5 (2007): 589–609.
8. See Kaushik Sunder Rajan, *Pharmocracy: Value, Politics, and Knowledge in Global Biomedicine* (Durham, NC: Duke University Press, 2017); Anne Roemer-Mahler, "Business Strategy and Access to Medicines in Developing Countries," *Global Health Governance* 4, no. 1 (2010), http://www.ghgj.org/Roemermahler4.1.htm.
9. "The State of Pharmaceutical Industry in India—an Overview," *Economic Times Healthworld*, August 29, 2017, accessed December 18, 2017, https://health.economictimes.indiatimes.com/news/pharma/the-state-of-pharmaceutical-industry-in-india-an-overview/60273583.
10. "On Closer Inspection: India's Drug Makers Have Come under Scrutiny" (by C.H.), *Schumpeter* (blog), *Economist*, February 20, 2014, accessed December 18, 2017, https://www.economist.com/blogs/schumpeter/2014/02/indias-booming-drugs-industry.
11. Naudé and Luiz, "Industry Analysis of Pharmaceutical Production in South Africa," 40.
12. Frederica Angeli, "With the Help of a Foreign Ally: Biopharmaceutical Innovation in India after TRIPS," *Health Policy and Planning* 29 (2014): 289.
13. Rajan, *Pharmocracy*, 6.
14. Pew Research Center, "Cell Phones in Africa: Communication Lifeline," April 2015, accessed October 27, 2015, http://www.pewglobal.org/files/2015/04/Pew-Research-Center-Africa-Cell-Phone-Report-FINAL-April-15-2015.pdf.
15. Toluwalogo Odumosu, "Making Mobiles African," in Mavhunga, *What Do Science, Technology, and Innovation Mean from Africa?*, 137–50.
16. Amitav Rath, "Science, Technology, and Policy at the Periphery: A View from the Centre," *World Development* 18, no. 11 (1990): esp. 1439.
17. Antina von Schnitzler, *Democracy's Infrastructure: Techno-politics and Protest after Apartheid* (Princeton, NJ: Princeton University Press, 2016), 196–97.
18. Suma Athreye and Andrew Godley, "Internationalising to Create Firm Specific Advantages: Leapfrogging Strategies of U.S. Pharmaceutical Firms in the 1930s and 1940s & Indian Pharmaceutical Firms in the 1990s and 2000s," United Nations University Working Paper Series 2008-051 (United Nations University—Maastricht Economic and Social Research and Training Centre on Innovation and Technology, Maastricht, the Netherlands, November 2007).
19. Skepticism of leapfrogging as an ideal has endured, however; see, e.g., Emmanuel M. Katongole, "Postmodern Illusions and the Challenges of African Theology," *Modern Theology* 16, no. 2 (2000): 237–54.
20. This idea is explored in the classic children's science fantasy book by Madeleine L'Engle, *A Wrinkle in Time* (New York: Farrar, Straus, and Giroux, 1962), esp. chap. 5, "The Tesseract."
21. An iThemba scientist who was involved in this endeavor, and had been sent to Cambridge for training, is now a professor at the University of Pretoria and has since been able to raise a combination of industry and university funds to import the flow chemistry technology. As of this writing this enterprise was making simple fertilizers rather than complicated pharmaceuticals.

22. This harnessing of biological re/productivity is fundamental to understandings of biovalue and biocapital. See Catherine Waldby, "Stem Cells, Tissue Cultures and the Production of Biovalue," *Health* 6, no. 3 (2002): 305–23; Rajan, *Biocapital.*
23. Marcus Baumann, Ian R. Baxendale, and Steven V. Ley, "The Flow Synthesis of Heterocycles for Natural Product and Medicinal Chemistry Applications," *Molecular Diversity* 15 (2011): 613.
24. Taylor, "Pharmaceutical Industry and Future of Drug Development," 15.
25. Solubility of intermediates becomes an important selection criterion: Mark D. Hopkin, Ian R. Baxendale, and Steven V. Ley, "An Expeditious Synthesis of Imatinib and Analogues Utilizing Flow Chemistry Methods," *Organic and Biomolecular Chemistry* 11, no. 11 (March 21, 2013): 1745–908.
26. Andrea Larson and Mark Meier, "The Business Case for Green Chemistry in the Pharmaceutical Industry," in *Green Techniques for Organic Synthesis and Medicinal Chemistry*, ed. Wei Zhang and Berkeley W. Cue Jr. (West Sussex, UK: John Wiley and Sons, 2012), 575.
27. "The pharmaceutical industry produces on the order of 10–1000 tons of product. In comparison, oil refining yields millions to hundreds of millions of tons of product. Nonetheless, the pharmaceutical industry may generate 25–100 tons of waste for every ton of product, whereas oil refining generates only 0.1 ton of waste per ton of product" (ibid.).
28. Elizabeth Grossman, *Chasing Molecules: Poisonous Products, Human Health, and the Promise of Green Chemistry* (Washington, DC: Shearwater, 2009), 13.
29. Concepcion Jimenez-Gonzalez et al., "Using the Right Green Yardstick: Why Process Mass Intensity Is Used in the Pharmaceutical Industry to Drive More Sustainable Processes," *Organic Process Research and Development* 15 (2011): 912–17.
30. James Michael Crow, "Going with the Flow: When It Comes to Scaling Up Organic Synthesis, It Pays to Think Small," *Chemistry World*, April 2012, 49.
31. Vahid Ebadat, "Dust Explosion Hazards in Pharmaceutical Facilities," Pharmaceutical Processing, October 24, 2012, accessed January 31, 2014, http://www.pharmpro.com/articles/2012/10/dust-explosion-hazards-pharmaceutical-facilities.
32. Indeed, there have since been even more portable flow chemistry systems developed—from shipping container sized to refrigerator sized; see "Going with the Flow," *Medicine Maker*, March 2017, 24–25, accessed December 18, 2017, https://themedicinemaker.com/issues/0317/going-with-the-flow/. However, for most purposes, the shipping container is a good module size.
33. Craig Martin, *Shipping Container* (New York: Bloomsbury, 2016).
34. Milne, "Pharmaceutical Prospects," 303.
35. This resonates with the modular approach taken by Doctors without Borders in their "global kit" for provision of medicine in emergency situations, as explored in Peter Redfield, *Life in Crisis: An Ethical Journey of Doctors without Borders* (Berkeley: University of California Press, 2013).
36. Kim Fortun, "essential2life," *Dialectical Anthropology* 34 (2010): 79.
37. Ibid., 81–85.
38. Edward Woodhouse and Steve Breyman, "Green Chemistry as a Social Movement?," *Science, Technology, and Human Values* 30, no. 2 (2005): 210.
39. Ibid., 203, 216.
40. Alex Scott, "Green Chemistry: Lighting Up New Pharma Pathways," *Chemical Week* 172, no. 6 (March 2010): 21. See also Alex Scott, "Pharma Manufacturers Chase Greener, Cost-Efficient Chemistries," *Chemical Week* 169, no. 5 (February 7, 2007): 21–28.

41. WHO, "Waste from Healthcare Activities," Fact Sheet no. 253, November 2011, accessed January 31, 2014, http://www.who.int/mediacentre/factsheets/fs253/en/.
42. Priyanka Singh, Bina Rani, and Raaz Maheshwari, "Pharmaceutical Pollution: A Short Communication," *International Journal of Pharmacy and Biological Sciences* 1, no. 2 (2011): 26–29.
43. Ranga Velagaleti and Philip Burns, "A Review of the Industrial Ecology of Particulate Pharmaceuticals and Waste Management Approaches," *Particulate Science and Technology: An International Journal* 25, no. 2 (2007): 120.
44. PricewaterhouseCoopers, "Pharma 2020: The Vision—Which Path Will You Take?," 2011, 9, accessed December 2, 2016, https://www.pwc.com/gx/en/industries/pharmaceuticals-life-sciences/publications/pharma-2020/pharma-2020-vision-path.html.
45. Larson and Meier, "Business Case for Green Chemistry," 575.
46. Joanna Chataway, Rebecca Hanlin, and Raphael Kaplinsky, "Inclusive Innovation: An Architecture for Policy Development," Innovation Knowledge Development Working Paper 65 (March 2013), esp. 10–11.
47. See Darren Brouwer, "Better Living through Chemistry?," *Comment*, June 23, 2016, accessed December 24, 2016, https://www.cardus.ca/comment/article/4895/better-living-through-chemistry.
48. Nikolas Rose and Carlos Novas, "Biological Citizenship," in Ong and Collier, *Global Assemblages*, 442.
49. Lisa Connock, "Ammonia and Methanol from Coal," *Nitrogen* 226 (March–April 1997): 47.
50. Clark, "Structured Inequality," 99.
51. Sparks, "Apartheid Modern," x.
52. Steven V. Ley, "On Being Green: Can Flow Chemistry Help?," *Chemical Record* 12 (2012): 378.
53. Mark D. Hopkin, Ian R. Baxendale, and Steven V. Ley, "A Flow-Based Synthesis of Imatinib: The API of Gleevec," *Chemical Communications* 46 (2010): 2450–52.
54. Pooja Van Dyck, "Importing Western Style, Exporting Tragedy: Changes in Indian Patent Law and Their Impact on AIDS Treatment in Africa," *Northwestern Journal of Technology and Intellectual Property* 6, no. 1 (Fall 2007): 138–51.
55. Tousignant, "Broken Tempos."
56. The profound untenability of this kind of nostalgia contrasts with that possible in places that had nationalist postcolonial "glory days" prior to fiscal austerity, such as Mozambique. See Ramah McKay, *Medicine in the Meantime: The Work of Care in Mozambique* (Durham, NC: Duke University Press, 2018), esp. a section on the "glory days of public health" (16–19).
57. Jasanoff and Kim, "Containing the Atom."
58. Ferguson, *Global Shadows*.
59. See Pollock, "Transforming the Critique of Big Pharma."
60. Médecins sans frontières Access Campaign, "Examples of the Importance of India as the 'Pharmacy of the Developing World,'" January 29, 2007, accessed February 8, 2014, https://www.msfaccess.org/content/examples-importance-india-pharmacy-developing-world; Unni Karunakara, "India: Double Frontal Attack on the 'Pharmacy of the Developing World,'" Médecins sans frontières, September 17, 2012, accessed November 30, 2015, https://www.msf.org/india-double-frontal-attack-pharmacy-developing-world.
61. Mbembe, *On the Postcolony*, 17.

EPILOGUE

1. I have explored the issue of pharmaceutical failure with regard to Big Pharma: Pollock, "Transforming the Critique of Big Pharma," 107.
2. Rajan, *Biocapital*, 129–30.
3. Mike Fortun, *Promising Genomics: Iceland and deCODE Genetics in a World of Speculation* (Berkeley: University of California Press, 2008), 196.
4. Noortje Marres and Linsey McGoey, "Experimental Failure: Notes on the Limits of the Performativity of Markets," 16–17, paper presented at "After Markets: Researching Hybrid Arrangements," Oxford Said Business School, UK, 2012, http://research.gold.ac.uk/7353/.
5. Drug Discovery and Development Center: H3D, accessed December 26, 2017, http://www.h3d.uct.ac.za.
6. "Vision and Mission," Drug Discovery and Development Center: H3D, accessed December 26, 2017, http://www.h3d.uct.ac.za/h3d-vision-and-mission-statement.
7. Nadia Krige, "Another Win for H3D as Kelly Chibale Achieves A-Rating," *University of Cape Town News*, December 20, 2017, accessed December 26, 2017, https://www.news.uct.ac.za/article/-2017-12-20-another-win-for-h3d-as-kelly-chibale-achieves-a-rating.
8. "UCT-H3D, MMV and International Partners Identify Second Potent Antimalarial Candidate," Medicines for Malaria Venture, July 27, 2016, accessed December 26, 2017, https://www.mmv.org/newsroom/press-releases/uct-h3d-mmv-and-international-partners-identify-second-potent-antimalarial.
9. "Flow Chemistry," University of Pretoria, accessed December 26, 2017, http://www.up.ac.za/en/chemistry/article/2581564/flow-chemistry.
10. "DRIVE: Infectious Innovations," DRIVE, accessed December 26, 2017, http://driveinnovations.org.
11. "Pharmaceutical Manufacturing (PMTC)—Limerick," Knowledge Transfer Ireland, accessed December 26, 2017, http://www.knowledgetransferireland.com/Research_in_Ireland/Research-Map-of-Ireland/HEI-Profiles/PMTC-Pharmaceutical-Manufacturing-Limerick.html.
12. Suman Seth, "Putting Knowledge in Its Place: Science, Colonialism, and the Postcolonial," *Postcolonial Studies* 12, no. 4 (2009): 373–78.
13. Matthew Harsh and James Smith, "Technology, Governance, and Place: Situating Biotechnology in Kenya," *Science and Public Policy* 34, no. 4 (May 2007): 251–60.
14. Robert E. Kohler, *Landscapes and Labscapes: Exploring the Lab-Field Border in Biology* (Chicago: University of Chicago Press, 2002).
15. David Arnold, *Colonizing the Body: State Medicine and Epidemic Disease in Nineteenth Century India* (Berkeley: University of California Press, 1993); Timothy Mitchell, *Colonizing Egypt* (Berkeley: University of California Press, 1988); Paul Rabinow, *French Modern: Norms and Forms of the Social Environment* (Cambridge, MA: MIT Press, 1989); Gwendolyn Wright, *The Politics of Design in French Colonial Urbanism* (Chicago: University of Chicago Press, 1991).
16. Brian Larkin, "The Politics and Poetics of Infrastructure," *Annual Review of Anthropology* 42 (2013): 329.
17. Susan Leigh Star, "The Ethnography of Infrastructure," *American Behavioral Scientist* 43, no. 3 (November/December 1999): 382.

BIBLIOGRAPHY

Abdool Karim, Salim S., and Quarraisha Abdool Karim. "AIDS Research Must Link to Local Policy." *Nature* 463 (February 2010): 733–34.

Abraham, Itty. "The Contradictory Spaces of Postcolonial Techno-science." *Economic and Political Weekly* 41, no. 3 (2006): 210–17.

Adichie, Chimamanda Ngozi. "The Danger of a Single Story." TED Talk, 2009. Accessed July 29, 2016. https://www.ted.com/talks/chimamanda_adichie_the_danger_of_a_single_story?language=en.

African Explosives and Chemical Industries. "History." Accessed January 18, 2016. http://www.aeci.co.za/aa_history.php.

Agnandji, Selidji T., Valerie Tsassa, Cornelia Conzelmann, Carsten Köhler, and Hans-Jörg Ehni. "Patterns of Biomedical Science Production in a Sub-Saharan Research Center." *BMC Medical Ethics* 13, no. 3 (March 2012): 1–7.

Akermann, Britt, and Faiz Kermani. "The Development of the South African Biotech Sector." *Journal of Commercial Biotech Sector* 12, no. 2 (2006): 111–19.

Al-Bader, Sara, Sarah E. Frew, Insiya Essagee, Victor Y. Liu, Abdallah S. Daar, and Peter A. Singer. "Small but Tenacious: South Africa's Biotech Sector." *Nature Biotechnology* 27, no. 5 (2009): 427–45.

Altona, R. E. "The Disposal of Industrial Effluents on Pastures." *Proceedings of the Annual Congresses of the Grassland Society of Southern Africa* 2, no. 1 (1967): 143–46.

Anderson, Warwick. "From Subjugated Knowledge to Conjugated Subjects: Science and Globalisation, or Postcolonial Studies of Science?" *Postcolonial Studies* 12, no. 4 (2009): 389–400.

———. "Postcolonial Technoscience." *Social Studies of Science* 32, nos. 4–5 (October–December 2002): 643–44.

Angeli, Frederica. "With the Help of a Foreign Ally: Biopharmaceutical Innovation in India after TRIPS." *Health Policy and Planning* 29 (2014): 280–91.

Appiah, Kwame Anthony. *Cosmopolitanism: Ethics in a World of Strangers*. New York: Norton, 2010.

Archer, Stephen L. "The Making of a Physician-Scientist—the Process Has a Pattern: Lessons from the Lives of Nobel Laureates in Medicine and Physiology." *European Heart Journal* 28 (2007): 510–14.

Arnold, David. *Colonizing the Body: State Medicine and Epidemic Disease in Nineteenth Century India*. Berkeley: University of California Press, 1993.

Ashforth, Adam. "Muthi, Medicine and Witchcraft: Regulating 'African Science' in Post-apartheid South Africa?" *Social Dynamics* 31, no. 2 (2005): 211–42.

Athreye, Suma, and Andrew Godley. "Internationalising to Create Firm Specific Advantages: Leapfrogging Strategies of U.S. Pharmaceutical Firms in the 1930s and 1940s & Indian Pharmaceutical Firms in the 1990s and 2000s." United Nations University Working Paper Series 2008-051, United Nations University—Maastricht Economic and Social Research and Training Centre on Innovation and Technology, Maastricht, the Netherlands, November 2007.

Barry, Andrew. "Pharmaceutical Matters: The Invention of Informed Materials." *Theory, Culture and Society* 22, no. 1 (2005): 51–69.

Bate, Roger, Ginger Zhe Jin, Aparna Mathur, and Amir Attaran. "Poor Quality Drugs in Global Trade: A Pilot Study." NBER Working Paper 20469, September 2014.

Baumann, Marcus, Ian R. Baxendale, and Steven V. Ley. "The Flow Synthesis of Heterocycles for Natural Product and Medicinal Chemistry Applications." *Molecular Diversity* 15 (2011): 613–30.

Beavon, Keith S. O. "Northern Johannesburg: Part of the 'Rainbow' or Neo-apartheid City in the Making?" *Mots pluriels* 13 (2000). Accessed June 25, 2016. http://motspluriels.arts.uwa.edu.au/MP1300kb.html.

Beer, John J. "Coal Tar Dye Manufacture and the Origins of the Modern Industrial Research Laboratory." *Isis* 49, no. 2 (1958): 123–31.

Behrens, Joanna. "The Dynamite Factory: An Industrial Landscape in Late-Nineteenth-Century South Africa." *Historical Archaeology* 39, no. 3 (2005): 61–74.

Bekker, Linda-Gail, Francois Venter, Karen Cohen, Eric Goemare, Gilles Van Cutsem, Andrew Boulle, and Robin Wood. "Provision of Antiretroviral Therapy in South Africa: The Nuts and Bolts." *Antiviral Therapy* 19, no. S3 (2014): 105–16.

Bell, Susan E., and Anne E. Figert. "Medicalization and Pharmaceuticalization at the Intersections: Looking Backward, Sideways and Forward." *Social Science and Medicine* 75 (2012): 2131–33.

Bell, Terry. "The Marikana Massacre: Why Heads Must Roll." *New Solutions: A Journal of Environmental and Occupational Health Policy* 25, no. 4 (2016): 440–50.

Beningfield, Jennifer. *The Freighted Land: Land, Landscape, and Politics in South Africa in the Twentieth Century*. London: Routledge, 2006.

Benjamin, Ruha. "A Lab of Their Own: Genomic Sovereignty as Postcolonial Science Policy." *Policy and Society* 28 (2009): 341–55.

Bennett, Jane. *Vibrant Matter: A Political Ecology of Things.* Durham, NC: Duke University Press, 2010.

Bensaude-Vincent, Bernadette. "Philosophy of Chemistry or Philosophy with Chemistry?" *HYLE: International Journal for the Philosophy of Chemistry* 20 (2014): 59–76.

Bensaude-Vincent, Bernadette, and Jonathan Simon. *Chemistry: The Impure Science.* London: Imperial College Press, 2008.

Bensaude-Vincent, Bernadette, and Isabelle Stengers. *A History of Chemistry*. Translated by Deborah van Dam. Cambridge, MA: Harvard University Press, 1996.

Bezuidenhout, Louise M., Sabina Leonelli, Ann H. Kelly, and Brian Rappert. "Beyond the Digital Divide: Towards a Situated Approach to Open Data." *Science and Public Policy* 44, no. 4 (2017): 464–75.

Bhabha, Homi. *The Location of Culture*. London: Routledge, 1994.

Biehl, João. "Pharmaceuticalization: AIDS Treatment and Global Health Politics." *Anthropological Quarterly* 80, no. 4 (2007): 1083–126.

Biesel, Uli, Ann H. Kelly, and Noémi Tousignant. "Knowing Insects: Hosts, Vectors, and the Companions of Science." *Science as Culture* 22, no. 1 (2013): 1–13.

Biruk, Crystal. *Cooking Data: Culture and Politics in an African Research World.* Durham, NC: Duke University Press, 2018.

Bond, Patrick. "Globalization, Pharmaceutical Pricing, and South African Health Policy: Managing Confrontation with US Firms and Politicians." *International Journal of Health Services* 29, no. 4 (October 1999): 765–92.

Bradley, Megan. "On the Agenda: North-South Research Partnerships and Agenda-Setting Practices." *Development in Practice* 18, no. 6 (November 2008): 673–85.

Brand South Africa. "South Africa Pays Less for ARV Drugs." December 15, 2010. Accessed December 23, 2017. https://www.brandsouthafrica.com/south-africa-fast-facts/health-facts/arvdrugs-151210.

Breckenridge, Keith. *Biometric State: The Global Politics of Identification and Surveillance in South Africa, 1850 to the Present.* Cambridge: Cambridge University Press, 2014.

———. "Marikana and the Limits of Biopolitics: Themes in the Recent Scholarship of South African Mining." *Africa* 84, no. 1 (February 2014): 151–61.

Brizuela-Garcia, Esperanza. "The History of Africanization and the Africanization of History." *History in Africa* 33 (2006): 85–100.

Brosman, Rocco. *Modderfontein Village Development Heritage Impact Assessment, Annexure 1: The History Report.* Johannesburg: Rocco Brosman, February 2010. Accessed December 26, 2017. https://modderconserve.files.wordpress.com/2010/05/modderfontein-hia-annexure-1-historical-report.pdf.

Brouwer, Darren. "Better Living through Chemistry?" *Comment,* June 23, 2016. Accessed December 24, 2016. https://www.cardus.ca/comment/article/4895/better-living-through-chemistry.

Brown, Stephen R. *A Most Damnable Invention: Dynamite, Nitrates, and the Making of the Modern World.* Toronto: Penguin Group, 2005.

Burton, Antoinette. *Africa in the Indian Imagination: Race and the Politics of Postcolonial Citation.* Durham, NC: Duke University Press, 2016.

Butler, Anthony, and Rosslyn Nicholson. *Life, Death and Nitric Oxide.* Cambridge: Royal Society of Chemistry, 2003.

Butler, Judith. *Bodies That Matter: On the Discursive Limits of Sex.* New York: Routledge, 1993.

Callon, Michel. "Some Elements of a Sociology of Translation: Domestication of the Scallops and the Fishermen of St. Brieuc Bay." In *Power, Action and Belief: A New Sociology of Knowledge,* edited by John Law, 196–233. London: Routledge and Kegan Paul, 1986.

"Carlos Slim Helu & Family." Profile. *Forbes.* https://www.forbes.com/profile/carlos-slim-helu/.

Carlson, David Gray. *A Commentary on Hegel's Science of Logic.* London: Palgrave Macmillan, 2007.

Cartwright, Alan Patrick. *The Dynamite Company: The Story of African Explosives and Chemical Industries.* Cape Town: Purnell and Sons, 1964.

Chambers, David Wade, and Richard Gillespie. "Locality in the History of Science: Colonial Science, Technoscience, and Indigenous Knowledge." *Osiris* 15 (2000): 221–40.

Chataway, Joanna, Rebecca Hanlin, and Raphael Kaplinsky. "Inclusive Innovation: An Architecture for Policy Development." Innovation Knowledge Development Working Paper 65, March 2013.

Cherry, Michael. "South African Science: Black, White, and Grey." *Nature* 463 (2010): 726–28.

Chetty, Nithaya, and Ahmed C. Bawa. "Physics for Development in Africa." *APS News* 14, no. 10 (November 2005): 8.

Chien, Colleen V. "HIV/AIDS Drugs for Sub-Saharan Africa: How Do Brand and Generic Supply Compare?" *PLoS ONE* 2, no. 3 (2007): e278. https://doi.org/10.1371/journal.pone.0000278.

Chinguno, Crispen. "Marikana Massacre and Strike Violence Post-partheid." *Global Labour Journal* 4, no. 2 (2013): 160–66.

Clark, Nancy L. *Manufacturing Apartheid: State Corporations in South Africa*. New Haven, CT: Yale University Press, 1994.

———. "Structured Inequality: Historical Realities of the Post-apartheid Economy." *Ufahamu: A Journal of African Studies* 38, no. 1 (2014): 93–118.

Cloatre, Emilie. *Pills for the Poorest: An Exploration of TRIPS and Access to Medication in Sub-Saharan Africa*. London: Palgrave Macmillan, 2013.

Cloatre, Emilie, and Robert Dingwall. "'Embedded Regulation': The Migration of Objects, Scripts, and Governance." *Regulation and Governance* 7 (2013): 365–86.

Cohen, Jon. "King of the Pills: Raymond Shinazi's Handful of Lifesaving Drugs Has Earned Him Riches, Esteem, and a Dose of Enmity." *Science* 348, no. 6235 (2015): 622–25.

Comaroff, Jean. "Beyond Bare Life: AIDS, (Bio)politics, and the Neoliberal Order." *Public Culture* 19, no. 1 (December 2007): 197–219.

———. "The Diseased Heart of Africa: Medicine, Colonialism, and the Black Body." In *Knowledge, Power, and Practice: The Anthropology of Medicine and Everyday Life*, edited by Shirley Lindenbaum and Margaret Lock, 305–29. Berkeley: University of California Press, 1993.

Comaroff, Jean, and John L. Comaroff. "Law and Disorder in the Postcolony: An Introduction." In *Law and Disorder in the Postcolony*, edited by Jean Comaroff and John L. Comaroff, 1–56. Chicago: University of Chicago Press, 2006.

———. *Theory from the South; or, How Euro-America Is Evolving toward Africa*. Boulder, CO: Paradigm, 2011.

Connock, Lisa. "Ammonia and Methanol from Coal." *Nitrogen* 226 (March–April 1997): 47.

Conservation Matters 5 (August–September 2017): 30. Accessed December 26, 2017. http://opus.sanbi.org/bitstream/123456789/5589/1/Magazine%20August%202017_final.pdf.

Cooper, Melinda. "On Pharmaceutical Empire: AIDS, Security, and Exorcism." In *Life as Surplus: Biotechnology and Capitalism in the Neoliberal Era*, 51–73. Seattle: University of Washington Press, 2008.

Cozzens, Susan. "Distributive Justice in Science and Technology Policy." *Science and Public Policy* 34, no. 2 (March 2007): 85–94.

Craddock, Susan. *Compound Solutions: Pharmaceutical Alternatives for Global Health*. Minneapolis: University of Minnesota Press, 2017.

———. "Drug Partnerships and Global Practices." *Health and Place* 18, no. 3 (2012): 481–89.

Crais, Clifton, and Thomas V. McClendon, eds. *The South Africa Reader: History, Culture, Politics*. Durham, NC: Duke University Press, 2013.

Crane, Johanna. "Adverse Events and Placebo Effects: African Scientists, HIV, and Ethics in the 'Global Health Sciences.'" *Social Studies of Science* 40, no. 6 (2010): 843–70.

———. *Scrambling for Africa: AIDS, Expertise, and the Rise of American Global Health Science*. Ithaca, NY: Cornell University Press, 2013.

Crankshaw, Owen. "Race, Space, and the Post-Fordist Spatial Order of Johannesburg." *Urban Studies* 45, no. 8 (July 2008): 1692–711.

Crow, James Michael. "Going with the Flow: When It Comes to Scaling Up Organic Synthesis, It Pays to Think Small." *Chemistry World*, April 2012, 48–51.

Daniel, John. "Soldiering On: The Post-presidential Years of Nelson Mandela." In *Legacies of Power: Leadership Change and Former Presidents in African Politics*, edited by Henning Melber and Roger Southall, 26–50. Cape Town: HSRC Press, 2006.

Das, Veena, and Ranendra K. Das. "Pharmaceuticals in Urban Ecologies: The Register of the Local." In *Global Pharmaceuticals: Ethics, Markets, Practices*, edited by Adriana Petryna, Andrew Lakoff, and Arthur Kleinman, 171–205. Durham, NC: Duke University Press, 2006.

Davie, Lucille. "Peaceful Park Hides Explosive Past." 2005. Accessed August 9, 2010. http://www.joburg.org.za/content/view/1109/168.

Droney, Damien. "Ironies of Laboratory Work during Ghana's Second Age of Optimism." *Cultural Anthropology* 29, no. 2 (2014): 363–84.

———. "Scientific Capacity Building and the Ontologies of Herbal Medicine in Ghana." *Canadian Journal of African Studies* 50, no. 3 (2016): 437–54.

Dronsfield, Alan. "Pain Relief: From Coal Tar to Paracetamol." Royal Society for Chemistry: Education in Chemistry. July 2005. Accessed July 2, 2016. http://www.rsc.org/education/eic/issues/2005July/painrelief.asp.

Du Bois, W. E. B. *The Souls of Black Folk*. Chicago: A. C. McClurg, 1903.

Dubow, Saul A. *Commonwealth of Knowledge: Science, Sensibility and White South Africa, 1820–2000*. Oxford: Oxford University Press, 2006.

———. "South Africa and South Africans: Nationality, Belonging, Citizenship." In *The Cambridge History of South Africa*, vol. 2, *1885–1994*, edited by Robert Ross, Anne Kelk Mager, and Bill Nasson, 17–65. Cambridge: Cambridge University Press, 2011.

Dumit, Joseph. *Drugs for Life: How Pharmaceutical Companies Define Our Health*. Durham, NC: Duke University Press, 2012.

Dyck, Pooja Van. "Importing Western Style, Exporting Tragedy: Changes in Indian Patent Law and Their Impact on AIDS Treatment in Africa." *Northwestern Journal of Technology and Intellectual Property* 6, no. 1 (Fall 2007): 138–51.

Ebadat, Vahid. "Dust Explosion Hazards in Pharmaceutical Facilities." Pharmaceutical Processing, October 24, 2012. Accessed January 31, 2014. http://www.pharmpro.com/articles/2012/10/dust-explosion-hazards-pharmaceutical-facilities.

Ecks, Stefan. "Pharmaceutical Citizenship: Antidepressant Marketing and the Promise of Demarginalization in India." *Anthropology and Medicine* 12, no. 3 (December 2005): 239–54.

Edlin, Chris. "The Importance of Patent Sharing in Neglected Disease Drug Discovery." *Future Medicinal Chemistry* 3, no. 11 (August 31, 2011): 1331–34.

Edlin, Chris D., Garreth Morgans, Susan Winks, Sandra Duffy, Vicky M. Avery, Sergio Wittlin, David Waterson, Jeremy Burrows, and Justin Bryans. "Identification and In-Vitro ADME Assessment of a Series of Novel Anti-malarial Agents Suitable for Hit-to-Lead Chemistry." *American Chemical Society Medicinal Chemistry Letters* 3, no. 7 (July 12, 2012): 570–73.

Edwards, Paul N., and Gabrielle Hecht. "History and the Technopolitics of Identity: The Case of Apartheid South Africa." *Journal of Southern African Studies* 36, no. 3 (September 2010): 619–39.

Erikson, Susan L. "Secrets from Whom? Following the Money in Global Health Finance." *Current Anthropology* 56, no. S12 (December 2015): S306–S316.

Evans, Martha. "The Last TV Star? Nelson Mandela's Funeral Broadcast, Social Media, and the Future of Media Events." In *Global Perspectives on Media Events in Contemporary Society*, edited by Andrew Fox, 141–57. Hershey, PA: Information Science Reference, 2016.

Farmer, Paul. *Infections and Inequalities: The Modern Plagues*. Berkeley: University of California Press, 1999.

Fassin, Didier. *When Bodies Remember: Experiences and Politics of AIDS in South Africa*. Berkeley: University of California Press, 2007.

Ferguson, James. *Expectations of Modernity: Myths and Meanings of Urban Life on the Zambian Copperbelt*. Berkeley: University of California Press, 1999.

———. *Global Shadows: Africa in the Neoliberal World Order*. Durham, NC: Duke University Press, 2006.

Fine, Ben, and Zavareh Rustomjee. *The Political Economy of South Africa: From the Minerals-Energy Complex to Industrialization*. Boulder, CO: Westview Press, 1996.

Fischer, Michael M. J. "Lively Capital and Translational Research." In *Lively Capital: Biotechnologies, Ethics, and Governance in Global Markets*, edited by Kaushik Sunder Rajan, 385–436. Durham, NC: Duke University Press, 2012.

Fisher, William W., III, and Cyrill P. Rigamonti. "The South Africa AIDS Controversy: A Case Study in Patent Law and Policy." Working Paper, Harvard Law School, February 10, 2005. Accessed January 31, 2016. http://cyber.law.harvard.edu/people/tfisher/South%20Africa.pdf.

Foley, Vernard L., and Patrick F. Belcastro. "William Brockedon and the Mechanization of Pill and Tablet Manufacture: From Bullets to Pills." *Pharmaceutical Technology*, September 1987, 110–16.

Fortun, Kim. *Advocacy after Bhopal: Environmentalism, Disaster, New Global Orders*. Chicago: University of Chicago Press, 2001.

———. "essential2life." *Dialectical Anthropology* 34 (2010): 77–86.

Fortun, Mike. *Promising Genomics: Iceland and deCODE Genetics in a World of Speculation*. Berkeley: University of California Press, 2008.

Foster, Laura A. "Decolonizing Patent Law: Postcolonial Technoscience and Indigenous Knowledge in South Africa." *Feminist Formations* 28, no. 3 (Winter 2016): 148–73.

———. "Inventing Hoodia: Vulnerabilities and Epistemic Citizenship in Southern Africa." *CSW Update*, April 2011, 15–19.

———. "A Postapartheid Genome: Genetic Ancestry Testing and Belonging in South Africa." *Science, Technology, and Human Values* 41, no. 6 (2016): 1015–36.

———. *Reinventing Hoodia: Peoples, Plants, and Patents in South Africa*. Seattle: University of Washington Press, 2017.

Freire, Paolo. *Pedagogy of Hope: Reliving "Pedagogy of the Oppressed."* Translated by Robert R. Barr. London: Bloomsbury Academic, 2014.

Friedman, Thomas. *The World Is Flat: A Brief History of the Twenty-First Century*. New York: Farrar, Straus, and Giroux, 2005.

Geissler, P. Wenzel, ed. *Para-states and Medical Science: Making African Global Health*. Durham, NC: Duke University Press, 2015.

———. "Public Secrets in Public Health: Knowing Not to Know When Making Scientific Knowledge." *American Ethnologist* 40, no. 1 (2013): 13–34.

Geissler, P. Wenzel, and Catherine Molyneaux. *Evidence, Ethos, and Experiment: The Anthropology and History of Medical Research in Africa*. Oxford: Berghahn Books, 2011.

Geissler, P. Wenzel, and Ruth Prince. "Active Compounds and Atoms of Society: Plants,

Bodies, Minds, and Cultures in the Work of Kenyan Ethnobotanical Knowledge." *Social Studies of Science* 39, no. 4 (2009): 599–634.

Gibson, Heather J., Matthew Walker, Brijesh Thapa, Kyriaki Kaplanidou, Sue Geldenhuys, and Willie Coetzee. "Psychic Income and Social Capital among Host Nation Residents: A Pre–Post Analysis of the 2010 FIFA World Cup in South Africa." *Tourism Management* 44 (2014): 113–22.

Gillespie, Kelly. "Reclaiming Nonracialism: Reading the Threat of Race from South Africa." *Patterns of Prejudice* 44, no. 1 (2010): 61–75.

Global Forum for Health Research. *The 10/90 Report on Health Research 2000*. Geneva: World Health Organization, 2000. Accessed December 26, 2017. http://announcementsfiles.cohred.org/gfhr_pub/assoc/s14791e/s14791e.pdf.

"Going with the Flow." *Medicine Maker*, March 2017, 24–25. Accessed December 18, 2017. https://themedicinemaker.com/issues/0317/going-with-the-flow/.

Goldberg, David Theo. *The Threat of Race: Reflections on Racial Neoliberalism*. Malden, MA: Wiley Blackwell, 2009.

Greene, Jeremy A. "Can the Government Stop the Next Martin Shkreli?" *Slate*, March 22, 2016. Accessed March 30, 2016. http://www.slate.com/articles/business/moneybox/2016/03/the_fda_wants_to_stop_the_next_martin_shkreli_by_speeding_up_the_approval.html.

———. *Generic: The Unbranding of Modern Medicine*. Baltimore: Johns Hopkins University Press, 2014.

———. *Prescribing by Numbers: Drugs and the Definition of Disease*. Baltimore: Johns Hopkins University Press, 2007.

Greenslit, Nathan. "Dep®ession and Comsum♀tion: Psychopharmaceuticals, Branding, and New Identity Practices." *Culture, Medicine and Psychiatry* 29, no. 4 (December 2005): 477–502.

Grossman, Elizabeth. *Chasing Molecules: Poisonous Products, Human Health, and the Promise of Green Chemistry*. Washington, DC: Shearwater, 2009.

Gumede, William. Introduction to *No Easy Walk to Freedom*, by Nelson Mandela, 7–28. Paarl, RSA: Kwela Books, 2013.

Gusterson, Hugh. "Nuclear Weapons and the Other in the Western Imagination." *Cultural Anthropology* 14, no. 1 (February 1999): 111–43.

———. "Studying Up Revisited." *PoLAR: Political and Legal Anthropology Review* 20, no. 1 (1997): 114–19.

Hall, Amy Laura. "Whose Progress? The Language of Global Health." *Journal of Medicine and Philosophy* 31 (2006): 285–304.

Haraway, Donna. "A Cyborg Manifesto: Science, Technology, and Socialist-Feminism in the Late Twentieth Century." In *Simians, Cyborgs, and Women: The Reinvention of Nature*, 149–81. New York: Routledge, 1991.

———. *Primate Visions: Gender, Race, and Nature in the World of Modern Science*. New York: Routledge, 1990.

———. "Situated Knowledges: The Science Question in Feminism and the Privilege of Partial Perspective." *Feminist Studies* 14, no. 3 (1988): 575–99.

Harding, Sandra. "Postcolonial and Feminist Philosophies of Science and Technology: Convergences and Dissonances." *Postcolonial Studies* 12, no. 4 (2009): 401–21.

———, ed. *The Postcolonial Science and Technology Studies Reader*. Durham, NC: Duke University Press, 2011.

Harsh, Matthew, and James Smith. "Technology, Governance, and Place: Situating Biotechnology in Kenya." *Science and Public Policy* 34, no. 4 (May 2007): 251–60.

Hayden, Cori. "A Generic Solution? Pharmaceuticals and the Politics of the Similar in Mexico." *Current Anthropology* 48, no. 4 (2007): 475–95.

———. "The Proper Copy: The Insides and Outsides of Domains Made Public." *Journal of Cultural Economy* 3, no. 1 (March 2010): 85–102.

———. "Rethinking Reductionism; or, The Transformative Work of Making the Same." *Anthropological Forum* 22, no. 3 (November 2012): 271–83.

———. *When Nature Goes Public: The Making and Unmaking of Bioprospecting in Mexico.* Princeton, NJ: Princeton University Press, 2003.

Hecht, Gabrielle. *Being Nuclear: Africans and the Global Uranium Trade.* Cambridge: Cambridge University Press, 2012.

Hokkanen, Markku. "Imperial Networks, Colonial Bioprospecting and Burroughs Wellcome & Co.: The Case of Strophanthus Kombe from Malawi (1859–1915)." *Social History of Medicine* 25, no. 3 (2012): 589–607.

Holmes, Linda Culp, and Frederick J. DiCarlo. "Nitroglycerin: The Explosive Drug." *Journal of Chemical Education* 48, no. 9 (1971): 573–76.

Hopkin, Mark D., Ian R. Baxendale, and Steven V. Ley. "An Expeditious Synthesis of Imatinib and Analogues Utilizing Flow Chemistry Methods." *Organic and Biomolecular Chemistry* 11, no. 11 (March 21, 2013): 1745–908.

———. "A Flow-Based Synthesis of Imatinib: The API of Gleevec." *Chemical Communications* 46 (2010): 2450–52.

Hughs, J. P., S. Rees, S. B. Kalindijan, and K. L. Philpott. "Principles of Early Drug Discovery." *British Journal of Pharmacology* 162 (2011): 1239–49.

Jackson, William A. "From Electuaries to Enteric Coating: A Brief History of Dosage Forms." In *Making Medicines: A Brief History of Pharmacy and Pharmaceuticals*, edited by Stuart Anderson, 203–22. London: Pharmaceutical Press, 2005.

Jasanoff, Sheila. "Future Imperfect: Science, Technology, and the Imaginations of Modernity." In *Dreamscapes of Modernity: Sociotechnical Imaginaries and the Fabrication of Power*, edited by Sheila Jasanoff and Sang-Hyun Kim, 1–33. Chicago: University of Chicago Press, 2015.

Jasanoff, Sheila, and Sang-Hyun Kim. "Containing the Atom: Sociotechnical Imaginaries and Nuclear Power in the United States and South Korea." *Minerva* 47, no. 2 (2009): 119–46.

———, eds. *Dreamscapes of Modernity: Sociotechnical Imaginaries and the Fabrication of Power*. Chicago: University of Chicago Press, 2015.

Jimenez-Gonzalez, Concepcion, Celia S. Ponder, Quirinus B. Broxterman, and Julie Manley. "Using the Right Green Yardstick: Why Process Mass Intensity Is Used in the Pharmaceutical Industry to Drive More Sustainable Processes." *Organic Process Research and Development* 15 (2011): 912–17.

Jones, Stuart, and André Müller. *The South African Economy, 1910–1990*. New York: St. Martin's Press, 1992.

Kale, Dinar, and Steve Little. "From Imitation to Innovation: The Evolution of R&D Capabilities and Learning Processes in the Indian Pharmaceutical Industry." *Technology Analysis and Strategic Management* 19, no. 5 (2007): 589–609.

Karunakara, Unni. "India: Double Frontal Attack on the 'Pharmacy of the Developing World.'" Médecins sans frontières, September 17, 2012. Accessed November 30, 2015. http://www.msf.org/india-double-frontal-attack-pharmacy-developing-world.

Katongole, Emmanuel M. "Postmodern Illusions and the Challenges of African Theology." *Modern Theology* 16, no. 2 (2000): 237–54.

Keller, Evelyn Fox. *A Feeling for the Organism: The Life and Work of Barbara McClintock*. San Francisco: W. H. Freeman, 1983.

———. *Reflections on Gender and Science*. New Haven, CT: Yale University Press, 1985.

Kelly, Ann H., David Ameh, Silas Majambere, Steve Lindsay, and Margaret Pinder. "'Like Sugar and Honey': Embedded Ethics of a Larval Control Project in The Gambia." *Social Science and Medicine* 70, no. 12 (2010): 1912–19.

Kelly, Ann H., and P. Wenzel Geissler, eds. *The Value of Transnational Medical Research: Labour, Participation, and Care*. London: Routledge, 2012.

Kock, Leon de. "South Africa in the Global Imaginary: An Introduction." *Poetics Today* 22, no. 2 (2001): 263–98.

Kohler, Robert E. *Landscapes and Labscapes: Exploring the Lab-Field Border in Biology*. Chicago: University of Chicago Press, 2002.

Kottakkunnummal, Manaf. "The Social Life of Indian Generic Pharmaceuticals in Johannesburg." PhD diss., University of the Witwatersrand, 2016.

Krige, Nadia. "Another Win for H3D as Kelly Chibale Achieves A-Rating." *University of Cape Town News*, December 20, 2017. Accessed December 26, 2017. https://www.news.uct.ac.za/article/-2017-12-20-another-win-for-h3d-as-kelly-chibale-achieves-a-rating.

Kuo, Wen Hua. "Understanding Race at the Frontier of Pharmaceutical Regulation: An Analysis of the Racial Difference Debate at the ICH." *Journal of Law, Medicine, and Ethics* 36, no. 3 (September 2008): 498–505.

Langwick, Stacey. *Bodies, Politics, and African Healing: The Matter of Maladies in Tanzania*. Bloomington: Indiana University Press, 2011.

Larkin, Brian. "The Politics and Poetics of Infrastructure." *Annual Review of Anthropology* 42 (2013): 327–43.

Larson, Andrea, and Mark Meier. "The Business Case for Green Chemistry in the Pharmaceutical Industry." In *Green Techniques for Organic Synthesis and Medicinal Chemistry*, edited by Wei Zhang and Berkeley W. Cue Jr., 573–87. West Sussex, UK: John Wiley and Sons, 2012.

Latour, Bruno. *Aramis; or, The Love of Technology*. Cambridge, MA: Harvard University Press, 1996.

———. "Give Me a Laboratory and I Will Raise the World." In *Science Observed: Perspectives on the Social Study of Science*, edited by Karin D. Knorr-Cetina and Michael Mulkay, 141–69. London: Sage, 1983.

Lemanski, Charlotte, Karina Landman, and Matthew Durington. "Divergent and Similar Experiences of 'Gating' in South Africa: Johannesburg, Durban and Cape Town." *Urban Forum* 19, no. 2 (2008): 133–58.

L'Engle, Madeleine. *A Wrinkle in Time*. New York: Farrar, Straus, and Giroux, 1962.

Leslie, Esther. *Nature, Art and the Chemical Industry*. London: Reaktion Books, 2005.

Ley, Steven V. "On Being Green: Can Flow Chemistry Help?" *Chemical Record* 12 (2012): 378–90.

Lezaun, Javier, and Catherine A. Montgomery. "The Pharmaceutical Commons: Sharing and Exclusion in Global Health Drug Development." *Science, Technology, and Human Values* 40, no. 1 (2015): 3–29.

Li, Jie Jack. *Laughing Gas, Viagra, and Lipitor: The Human Stories behind the Drugs We Use*. New York: Oxford University Press, 2006.

Liebenau, Jonathan. "Ethical Business: The Formation of the Pharmaceutical Industry in Britain, Germany and the United States before 1914." *Business History* 30 (1988): 116–29.

Limb, Peter. "The Digitization of Africa." *Africa Today* 52, no. 2 (Winter 2005): 3–19.

Liotta, Dennis C., and George R. Painter. "Discovery and Development of the Anti-human Immunodeficiency Virus Drug, Emtricitabine (Emtriva, FTC)." *Accounts of Chemical Research* 49, no. 10 (2016): 2091–98.

Livingston, Julie. *Improvising Medicine: An African Oncology Ward in an Emerging Epidemic*. Durham, NC: Duke University Press, 2012.

Lorde, Audre. "The Master's Tools Will Never Dismantle the Master's House." In *Sister Outsider: Essays and Speeches*, 110–13. Trumansburg, NY: Crossing Press, 1984.

Maldonado-Torres, Nelson. "On the Coloniality of Being: Contributions to the Development of a Concept." *Cultural Studies* 21, nos. 2–3 (2007): 240–70.

Mandela, Nelson. "Forward." In *Building a New South Africa*, vol. 3, *Science and Technology Policy: A Report from the Mission on Science and Technology Policy for a Democratic South Africa*, edited by Marc Van Ameringen, vii–viii. Ottawa, Canada: International Development Research Centre, 1995.

———. "Inaugural Address." In *The South Africa Reader: History, Culture, Politics*, ed. Clifton Crais and Thomas V. McClendon, 470–72. Durham, NC: Duke University Press, 2013.

Marks, Harry M. *The Progress of Experiment: Science and Therapeutic Reform in the United States, 1900–1990*. Cambridge: Cambridge University Press, 1997.

Marres, Noortje, and Linsey McGoey. "Experimental Failure: Notes on the Limits of the Performativity of Markets." Paper presented at "After Markets: Researching Hybrid Arrangements," Oxford Said Business School, UK, 2012. http://research.gold.ac.uk/7353/.

Martin, Craig. *Shipping Container*. New York: Bloomsbury, 2016.

Martin, Emily. "The Pharmaceutical Person." *BioSocieties* 1, no. 1 (2006): 273–87.

Marx, Karl. *Capital*. Vol. 3. 1894. Accessed October 2, 2016. https://www.marxists.org/archive/marx/works/download/pdf/Capital-Volume-III.pdf.

———. *The Poverty of Philosophy*. 1847. Translated by Harry Quelch. New York: Cosimo, 2008.

Matsinhe, David M. *Apartheid Vertigo: The Rise in Discrimination against Africans in South Africa*. Farnham, UK: Ashgate, 2011.

Mattingly, Cheryl. *The Paradox of Hope: Journeys through a Clinical Borderland*. Berkeley: University of California Press, 2010.

Mavhunga, Clapperton Chakanetsa. *What Do Science, Technology, and Innovation Mean from Africa?* Cambridge, MA: MIT Press, 2017.

Mbali, Mandisa. *South African AIDS Activism and Global Health Politics*. London: Palgrave Macmillan, 2013.

Mbeki, Thabo. "I Am an African." In *The South Africa Reader: History, Culture, Politics*, ed. Clifton Crais and Thomas V. McClendon, 475–80. Durham, NC: Duke University Press, 2013.

Mbembe, Achille. *On the Postcolony*. Berkeley: University of California Press, 2001.

McCall, Leslie. "The Complexity of Intersectionality." *Signs: Journal of Women in Culture and Society* 30, no. 3 (2005): 1771–1800.

McClendon, Thomas, and Pamela Scully. "The South African Student Exchange Program: Anti-apartheid Activism in the Era of Constructive Engagement." *Safundi: The Journal of South African and American Studies* 16, no. 1 (2015): 1–27.

McDowell, Andrew. "Making the Global Equivalent: Markets, Relations and Pharmaceuticals in the Anthropology of Global Health in Africa." *BioSocieties* 10, no. 3 (September 2015): 380–84.

McGoey, Linsey. *No Such Thing as a Free Gift: The Gates Foundation and the Price of Philanthropy*. London: Verso, 2015.

M'charek, Amade. "Race, Time and Folded Objects: The HeLa Error." *Theory Culture Society* 31, no. 6 (2014): 29–56.

McKay, Ramah. *Medicine in the Meantime: The Work of Care in Mozambique*. Durham, NC: Duke University Press, 2018.

McNeil, Donald G., Jr. "South Africa's Bitter Pill for World's Drug Makers." *New York Times*, March 29, 1998. http://www.nytimes.com/1998/03/29/business/south-africa-s-bitter-pill-for-world-s-drug-makers.html.

Médecins sans frontières Access Campaign. "Examples of the Importance of India as the 'Pharmacy of the Developing World.'" January 29, 2007. Accessed February 8, 2014. https://www.msfaccess.org/content/examples-importance-india-pharmacy-developing-world.

———. "1998: Big Pharma versus Nelson Mandela." January 2009. Accessed February 3, 2016. http://www.msfaccess.org/content/1998-big-pharma-versus-nelson-mandela.

Medicines for Malaria Venture. "UCT-H3D, MMV and International Partners Identify Second Potent Antimalarial Candidate." July 27, 2016. Accessed December 26, 2017. https://www.mmv.org/newsroom/press-releases/uct-h3d-mmv-and-international-partners-identify-second-potent-antimalarial.

Mika, Marissa. "Fifty Years of Creativity, Crisis, and Cancer in Uganda." *Canadian Journal of African Studies / Revue canadienne des études africaines* 50, no. 3 (2016): 395–413.

Milne, Richard. "Pharmaceutical Prospects: Biopharming and the Geography of Technological Expectations." *Social Studies of Science* 42, no. 2 (2012): 290–306.

Milner, Katie M., and Nancy Baym. "The Selfie of the Year of the Selfie: Reflections on a Media Scandal." *International Journal of Communication* 9 (2015): 1701–15.

Mitchell, Timothy. *Colonizing Egypt*. Berkeley: University of California Press, 1988.

Mkhwanazi, Nolwazi. "Medical Anthropology in Africa: The Trouble with a Single Story." *Medical Anthropology* 35, no. 2 (2016): 193–202.

Mohamed, Seeraj. "The Energy-Intensive Sector: Considering South Africa's Comparative Advantage in Cheap Energy." *Trade and Industrial Policy Strategies (TIPS) Forum*, 1998. Accessed June 18, 2016. http://www.tips.org.za/research-archive/annual-forum-papers/1998/item/download/8_6b0779ade5b600f6ce8e9ff3c110e022.

Mol, Annemarie, and John Law. "Regions, Networks and Fluids: Anaemia and Social Topology." *Social Studies of Science* 24, no. 4 (November 1994): 641–71.

Montgomery, Catherine M. "Making Prevention Public: The Co-production of Gender and Technology in HIV Prevention Research." *Social Studies of Science* 42, no. 6 (December 2012): 922–44.

Moore-Sheeley, Kirsten. "The Products of Experiment: Changing Conceptions of Difference in the History of Tuberculosis in East Africa, 1920s–1970s." *Social History of Medicine*, June 28, 2017. https://doi.org/10.1093/shm/hkx048.

Morris, Christopher. "Biopolitics and Boundary Work in South Africa's Sutherlandia Clinical Trial." *Medical Anthropology* 36, no. 7 (2017): 685–98.

Morris, Peter J. T. *The Matter Factory: A History of the Chemistry Laboratory*. London: Reaktion Books, 2015.

Mudimbe, V. Y. *The Invention of Africa: Gnosis, Philosophy, and the Order of Knowledge*. Bloomington: Indiana University Press, 1988.

Muldoon, Katherine A., Josephine Birungi, Nicole S. Berry, Moses H. Ngolobe, Robert Mwesigwa, Kate Shannon, and David M. Moore. "Supporting Southern-Led Research:

Implications for North-South Research Partnerships." *Canadian Journal of Public Health* 103, no. 2 (March 2012): 128–31.

Murphy, Michelle. "Distributed Reproduction, Chemical Violence, and Latency." *Scholar and Feminist Online* 11, no. 3 (Summer 2013).

Nachega, Jean B., Jean-Jacques Parienti, Olalekan A. Uthman, Robert Gross, David W. Dowdy, Paul E. Sax, Joel E. Gallant, Michael J. Mugavero, Edward J. Mills, and Thomas P. Giordano. "Lower Pill Burden and Once-Daily Dosing Antiretroviral Treatment Regimens for HIV Infection: A Meta-analysis of Randomized Controlled Trials." *Clinical Infectious Diseases* 58, no. 9 (2014): 1297–1307.

Nader, Laura. "Up the Anthropologist: Perspectives Gained from Studying Up." In *Revisiting Anthropology*, edited by Dell Hymes, 284–311. 1969; New York: Vintage Books, 1974.

Nakamura, Lisa. "Indigenous Circuits: Navajo Women and the Racialization of Early Electronic Manufacture." *American Quarterly* 66, no. 4 (December 2014): 919–41.

Nattrass, Nicoli. *Mortal Combat: AIDS Denialism and the Struggle for Antiretrovirals in South Africa*. Scottsville: University of KwaZulu-Natal Press, 2007.

Naudé, C. te W., and J. M. Luis. "An Industry Analysis of Pharmaceutical Production in South Africa." *South African Journal of Business Management* 44, no. 1 (2013): 33–46.

Nchinda, Thomas C. "Research Capacity Strengthening in the South." *Social Science and Medicine* 54, no. 11 (June 2002): 1699–711.

Ndlovu-Gatsheni, Sabelo J. "Pan-Africanism and the 2010 FIFA World Cup in South Africa." *Development Southern Africa* 28, no. 3 (September 2011): 401–13.

Negroponte, Nicholas. *Being Digital*. New York: Knopf, 1995.

———. "Bits and Atoms." January 1, 1995. Accessed June 30, 2016. http://web.media.mit.edu/~nicholas/Wired/WIRED3-01.html.

Neill, Deborah. "Paul Ehrlich's Colonial Connections: Scientific Networks and Sleeping Sickness Drug Therapy Research, 1900–1914." *Social History of Medicine* 22, no. 1 (2009): 61–77.

Nelson, Alondra. "The Inclusion-and-Difference Paradox." Review of *Inclusion: The Politics of Difference in Medical Research*, by Steven Epstein. *Social Identities* 15 (2009): 742–43.

Nelson, Alondra, and Thuy Linh N. Tu, eds. *TechniColor: Race, Technology, and Everyday Life*. New York: New York University Press, 2001.

Nguyen, Vinh-Kim. "Anti-retroviral Globalism, Biopolitics, and Therapeutic Citizenship." In *Global Assemblages: Technology, Politics, and Ethics as Anthropological Problems*, edited by Aihwa Ong and Stephen J. Collier, 124–44. Malden, MA: Blackwell, 2005.

Nixon, Rob. *Slow Violence and the Environmentalism of the Poor*. Cambridge, MA: Harvard University Press, 2011.

Nordling, Linda. "Made in Africa." *Nature Medicine* 19, no. 7 (2013): 803–6.

Nuttall, Sarah. "Subjectivities of Whiteness." *African Studies Review* 44, no. 2 (2001): 115–40.

Obama, Barack. Transcript of speech at the Democratic National Convention, July 27, 2004. Accessed December 30, 2016. http://www.librarian.net/dnc/speeches/obama.txt.

O'Brien, Kevin. "Drug Companies and AIDS in Africa." *America Magazine*, November 25, 2002. Accessed November 10, 2017. https://www.americamagazine.org/issue/413/article/drug-companies-and-aids-africa.

O'Connell, Siona. "A Search for the Human in the Shadow of Rhodes." *Ufahamu: A Journal of African Studies* 38, no. 3 (2015): 11–14.

Odumosu, Toluwalogo. "Making Mobiles African." In *What Do Science, Technology, and Innovation Mean from Africa?*, edited by Clapperton Chakanetsa Mavhunga, 137–50. Cambridge, MA: MIT Press, 2017.

Okeke, Iruka N. *Divining without Seeds: The Case for Strengthening Laboratory Medicine in Africa*. Ithaca, NY: Cornell University Press, 2011.

Okwaro, Ferdinand Moyi, and P. W. Geissler. "In/dependent Collaborations: Perceptions and Experiences of African Scientists in Transnational HIV Research." *Medical Anthropology Quarterly* 29, no. 4 (December 2015): 492–511.

"On Closer Inspection: India's Drug Makers Have Come under Scrutiny." (By C.H.) *Schumpeter* (blog), *Economist*, February 20, 2014. Accessed December 18, 2017. https://www.economist.com/blogs/schumpeter/2014/02/indias-booming-drugs-industry.

Osseo-Asare, Abena. "Bioprospecting and Resistance: Transforming Poisoned Arrows into Strophanthin Pills in Colonial Gold Coast, 1885–1922." *Social History of Medicine* 21, no. 2 (2008): 269–90.

———. *Bitter Roots: The Search for Healing Plants in Africa*. Chicago: University of Chicago Press, 2014.

Packard, Randall M. *White Plague, Black Labor: Tuberculosis and the Political Economy of Health and Disease in South Africa*. Berkeley: University of California Press, 1989.

Patel, Aarti, Robin Gauld, Pauline Norris, and Thomas Rades. "'This Body Does Not Want Free Medicines': South African Consumer Perceptions of Drug Quality." *Health Policy and Planning* 25 (2010): 61–69.

Patton, Cindy. "Inventing 'African AIDS.'" *new formations* 10 (Spring 1990): 25–39.

Persson, Asha. "Incorporating Pharmakon: HIV, Medicine, and Body Shape Change." *Body and Society* 10, no. 4 (2004): 45–67.

Peterson, Kristin. "AIDS Policies for Markets and Warriors: Dispossession, Capital, and Pharmaceuticals in Nigeria." In *Lively Capital: Biotechnologies, Ethics, and Governance in Global Markets*, edited by Kaushik Sunder Rajan, 228–47. Durham, NC: Duke University Press, 2012.

———. *Speculative Markets: Drug Circuits and Derivative Life in Nigeria*. Durham, NC: Duke University Press, 2014.

Petryna, Adriana. *When Experiments Travel: Clinical Trials and the Global Search for Human Subjects*. Princeton, NJ: Princeton University Press, 2009.

Pew Research Center. "Cell Phones in Africa: Communication Lifeline." April 2015. Accessed October 27, 2015. http://www.pewglobal.org/files/2015/04/Pew-Research-Center-Africa-Cell-Phone-Report-FINAL-April-15-2015.pdf.

Pharasi, Bada, and Jacqui Miot. "Medicines Selection and Procurement in South Africa." *South African Health Review*, 2013, 117–85.

Phflanz, Mike. "Nelson Mandela: The Quiet Street Where He Died Is Filled with Clapping and Singing." *Telegraph* (London), December 6, 2013. Accessed October 21, 2017. http://www.telegraph.co.uk/news/worldnews/nelson-mandela/10501578/Nelson-Mandela-the-quiet-street-where-he-died-is-filled-with-clapping-and-singing.html.

Pienaar, Kiran. "Claiming Rights, Making Citizens: HIV and the Performativity of Biological Citizenship." *Social Theory and Health* 14, no. 2 (May 2016): 149–68.

Pollack, Andrew. "Drug Goes from $13.50 a Tablet to $750, Overnight." *New York Times*, September 20, 2015. https://www.nytimes.com/2015/09/21/business/a-huge-overnight-increase-in-a-drugs-price-raises-protests.html.

Pollock, Anne. *Medicating Race: Heart Disease and Durable Preoccupations with Difference*. Durham, NC: Duke University Press, 2012.

———. "Pharmaceutical Meaning-Making eyond Marketing: Racialized Subjects of Generic Thiazide." *Journal of Law, Medicine, and Ethics* 36, no. 3 (2008): 530–36.

———. "Places of Pharmaceutical Knowledge-Making: Global Health, Postcolonial Science, and South African Drug Discovery." *Social Studies of Science* 44, no. 6 (December 2014): 848–73.

———. "Transforming the Critique of Big Pharma." *BioSocieties* 6, no. 1 (2011): 106–18.

Pollock, Anne, and David S. Jones. "Coronary Artery Disease and the Contours of Pharmaceuticalization." *Social Science and Medicine* 131 (April 2015): 221–27.

Pollock, Anne, and Banu Subramaniam. "Resisting Power, Retooling Justice: Promises of Feminist Postcolonial Technosciences." *Science, Technology, and Human Values* 41, no. 6 (2016): 951–66.

Posel, Deborah. "Sex, Death, and the Fate of the Nation: Reflections on the Politicization of Sexuality in Post-apartheid South Africa." *Africa: Journal of the International African Institute* 75, no. 2 (2005): 125–53.

Pottage, Alain. "Law Machines: Scale Models, Forensic Materiality and the Making of Modern Patent Law." *Social Studies of Science* 41, no. 5 (2011): 621–43.

Powell, Walter W., and Kaisa Snellman. "The Knowledge Economy." *Annual Review of Sociology* 30 (2004): 199–220.

PricewaterhouseCoopers. "Pharma 2020: The Vision—Which Path Will You Take?" 2011. Accessed December 2, 2016. https://www.pwc.com/gx/en/industries/pharmaceuticals-life-sciences/publications/pharma-2020/pharma-2020-vision-path.html.

Puig de la Bellacasa, Maria. "Matters of Care in Technoscience: Assembling Neglected Things." *Social Studies of Science* 41, no. 1 (2011): 85–106.

Rabinow, Paul. *French Modern: Norms and Forms of the Social Environment*. Cambridge, MA: MIT Press, 1989.

Race, Kane. *Pleasure Consuming Medicine: The Queer Politics of Drugs*. Durham, NC: Duke University Press, 2009.

Rajan, Kaushik Sunder. *Biocapital: The Constitution of Postgenomic Life*. Durham, NC: Duke University Press, 2006.

———. "Experimental Values: Indian Trials and Surplus Health." *New Left Review* 45 (2007): 67–88.

———. *Pharmocracy: Value, Politics, and Knowledge in Global Biomedicine*. Durham, NC: Duke University Press, 2017.

Rath, Amitav. "Science, Technology, and Policy at the Periphery: A View from the Centre." *World Development* 18, no. 11 (1990): 1429–43.

Redfield, Peter. "Bioexpectations: Life Technologies as Humanitarian Goods." *Public Culture* 24, no. 1 (2012): 157–84.

———. *Life in Crisis: An Ethical Journey of Doctors without Borders*. Berkeley: University of California Press, 2013.

Reihling, Hanspeter C. W. "Bioprospecting the African Renaissance: The New Value of *Muthi* in South Africa." *Journal of Ethnobiology and Ethnomedicine* 4, no. 9 (2008). Accessed October 10, 2017. http://www.ethnobiomed.com/content/4/1/9.

Ringertz, Nils. "Alfred Nobel—His Life and Work." *Nature Reviews Molecular Cell Biology* 2, no. 12 (December 2001): 925–28.

Ripin, David J., David Jamieson, Amy Meyers, Umesh Warty, Mary Dain, and Cyril Khamsi. "Antiretroviral Procurement and Supply Chain Management." *Antiviral Therapy* 19, no. S3 (2014): 79–89.

Robins, Steven. "From 'Rights' to 'Ritual': AIDS Activism in South Africa." *American Anthropologist* 108, no. 2 (June 2006): 312–33.

———. "'Long Live Zackie, Long Live': AIDS Activism, Science and Citizenship after Apartheid." *Journal of Southern African Studies* 30, no. 3 (September 2004): 651–72.

Roemer-Mahler, Anne. "Business Strategy and Access to Medicines in Developing Countries." *Global Health Governance* 4, no. 1 (2010). http://www.ghgj.org/Roemermahler 4.1.htm.

Rose, Hilary. "Hand, Brain, and Heart: A Feminist Epistemology for the Natural Sciences." *Signs* 9, no. 1 (Autumn 1983): 73–90.

Rose, Nikolas. *The Politics of Life Itself: Biomedicine, Power, and Subjectivity in the Twenty-First Century.* Princeton, NJ: Princeton University Press, 2006.

Rose, Nikolas, and Carlos Novas. "Biological Citizenship." In *Global Assemblages: Technology, Politics, and Ethics as Anthropological Problems*, edited by Aihwa Ong and Stephen J. Collier, 439–63. Malden, MA: Blackwell, 2005.

Rosengarten, Marsha. *HIV Interventions: Biomedicine and the Traffic between Information and Flesh.* Seattle: University of Washington Press, 2009.

Rosenthal, Joshua. "Deconstructing a Research Project." Review of *When Nature Goes Public*, by Cori Hayden. *EMBO Reports* 5, no. 11 (2004): 1035–36.

Rottenburg, Richard. "Social and Public Experiments and New Figurations of Science and Politics in Postcolonial Africa." *Postcolonial Studies* 12, no. 4 (2009): 423–40.

Rovira, Joan. "Creating and Promoting Domestic Drug Manufacturing Capacities: A Solution for Developing Countries?" In *Negotiating Health: Intellectual Property and Access to Medicines*, edited by Pedro Roffe, Geoff Tansey, and David Vivas-Eugui, 227–40. London: Earthscan, 2006.

Roy, Deboleena. "Germline Ruptures: Methyl Isocyanate Gas and the Transpositions of Life, Death, and Matter in Bhopal." Paper presented at University of California, Los Angeles, Center for the Study of Women, Life (Un)Ltd. Series, November 2013.

Saethre, Erik, and Jonathan Stadler. *Negotiating Pharmaceutical Uncertainty: Women's Agency in a South African HIV Prevention Trial.* Nashville, TN: Vanderbilt University Press, 2017.

Said, Edward. *Orientalism.* New York: Pantheon Books, 1978.

Saunders, Christopher. "Perspective on the Transition from Apartheid to Democracy in South Africa." *South African Historical Journal* 51, no. 1 (2004): 159–66.

Schneider, Helen. "On the Fault-Line: The Politics of AIDS Policy in Contemporary South Africa." *African Studies* 61, no. 1 (2002): 145–67.

Schnitzler, Antina von. *Democracy's Infrastructure: Techno-politics and Protest after Apartheid.* Princeton, NJ: Princeton University Press, 2016.

———. "Traveling Technologies: Infrastructure, Ethical Regimes, and the Materiality of Politics in South Africa." *Cultural Anthropology* 28, no. 4 (2013): 670–93.

Schumaker, Lyn. *Africanizing Anthropology: Fieldwork, Networks, and the Making of Cultural Knowledge in Central Africa.* Durham, NC: Duke University Press, 2001.

Scott, Alex. "Green Chemistry: Lighting Up New Pharma Pathways." *Chemical Week* 172, no. 6 (March 2010): 21–23.

———. "Pharma Manufacturers Chase Greener, Cost-Efficient Chemistries." *Chemical Week* 169, no. 5 (February 7, 2007): 21–28.

Scott-Burden, Timothy. "Nitric Oxide Leads to Prized NObility: Background to the Work of Ferid Murad." *Texas Heart Journal* 26, no. 1 (1999): 1–5.

Šehović, Annamarie Bindenagel. *HIV/AIDS and the South African State: Sovereignty and the Responsibility to Respond.* Farnham, UK: Ashgate, 2014.

Sekhejane, Palesa, and Charlotte Pelletan. "HIV and AIDS Triumphs and Struggles." In *Sizonqoba! Outliving AIDS in Southern Africa*, edited by Busani Ngcaweni, 94–115. Pretoria: Africa Institute of South Africa, 2016.

Serres, Michel, with Bruno Latour. *Conversations on Science, Culture, and Time*. Ann Arbor: University of Michigan Press, 1995.

Seth, Suman. "Putting Knowledge in Its Place: Science, Colonialism, and the Postcolonial." *Postcolonial Studies* 12, no. 4 (2009): 373–78.

Sethi, S. Prakash, and Oliver F. Williams. *Economic Imperatives and Ethical Values in Global Business: The South African Experience and International Codes Today*. Boston: Kluwer Academic, 2000.

Shaloff, Stanley. "The Africanization Controversy in the Gold Coast, 1926–1946." *African Studies Review* 17, no. 3 (December 1974): 493–504.

Singh, Priyanka, Bina Rani, and Raaz Maheshwari. "Pharmaceutical Pollution: A Short Communication." *International Journal of Pharmacy and Biological Sciences* 1, no. 2 (2011): 26–29.

Slinn, Judy. "The Development of the Pharmaceutical Industry." In *Making Medicines: A Brief History of Pharmacy and Pharmaceuticals*, edited by Stuart Anderson, 155–74. London: Pharmaceutical Press, 2005.

Sonopo, M. S., K. Venter, G. Boyle, S. Winks, B. Marjanovic-Painter, and J. R. Zeevaart. "Carbon-14 Radiolabeling and In Vivo Biodistribution of a Potential Anti-TB Compound." *Journal of Labelled Compounds and Radiopharmaceuticals* 58, no. 2 (February 2015): 23–29.

Sooryamoorthy, Radhamany. "Science and Scientific Collaboration in South Africa: Apartheid and After." *Scientometrics* 84 (2010): 373–90.

"South Africa's Opportunity." *Nature* 463 (February 2010): 709.

Sparks, Stephen John. "Apartheid Modern: South Africa's Oil from Coal Project and the History of a South African Company Town." PhD diss., University of Michigan, 2012.

Spivak, Gayatri Chakravorty. "Can the Subaltern Speak?" (abbreviated by the author). In *The Post-colonial Studies Reader*, edited by Bill Ashcroft, Gareth Griffiths, and Helen Tiffin, 28–37. New York: Routledge, 1995.

Star, Susan Leigh. "The Ethnography of Infrastructure." *American Behavioral Scientist* 43, no. 3 (November/December 1999): 377–91.

"The State of Pharmaceutical Industry in India—an Overview." *Economic Times Healthworld*, August 29, 2017. Accessed December 18, 2017. https://health.economictimes.indiatimes.com/news/pharma/the-state-of-pharmaceutical-industry-in-india-an-overview/60273583.

Statistics South Africa. "Mortality and Causes of Death in South Africa, 2013: Findings from Death Notification." Accessed September 17, 2015. http://www.statssa.gov.za/publications/P03093/P030932013.pdf.

Steenkamp, W. F. J. "The Pharmaceutical Industry in South Africa." *South African Journal of Economics* 47, no. 1 (1979): 46–57.

Stoler, Ann Laura. "'The Rot Remains': From Ruins to Ruination." In *Imperial Debris: On Ruins and Ruination*, edited by Ann Laura Stoler, 1–38. Durham, NC: Duke University Press, 2013.

Storey, William Kelleher. "Cecil Rhodes and the Making of a Sociotechnical Imaginary for South Africa." In *Dreamscapes of Modernity: Sociotechnical Imaginaries and the Fabrication of Power*, edited by Sheila Jasanoff and Sang-Hyun Kim, 34–55. Chicago: University of Chicago Press, 2015.

———. *Guns, Race, and Power in Colonial South Africa*. New York: Cambridge University Press, 2008.

Sturtevant, William C. "Studies in Ethnoscience." *American Anthropologist* 66, no. 3, pt. 2 (June 1964): 99–131.

Subramaniam, Banu. "Colonial Legacies, Postcolonial Biologies: Gender and the Promises of Biotechnology." *Asian Biotechnology and Development Review* 17, no. 1 (March 2015): 15–36.

———. *Ghost Stories for Darwin: The Science of Variation and the Politics of Diversity.* Urbana: University of Illinois Press, 2014.

TallBear, Kim. *Native American DNA: Tribal Belonging and the False Promise of Genetic Science.* Minneapolis: University of Minnesota Press, 2013.

Taylor, David. "The Pharmaceutical Industry and the Future of Drug Development." In *Pharmaceuticals in the Environment,* edited by R. E. Hester and R. M. Harrison, 1–33. London: Royal Society of Chemistry, 2015.

Terry, Fiona. *Condemned to Repeat? The Paradox of Humanitarian Action.* Ithaca, NY: Cornell University Press, 2002.

Tilley, Helen. *Africa as a Living Laboratory: Empire, Development, and the Problem of Scientific Knowledge, 1870–1950.* Chicago: University of Chicago Press, 2011.

Torsoli, Albertina, and Mikiko Kitamura. "Drug Makers Join Gates Foundation to Halt Tropical Illness." Bloomberg, January 30, 2012. Accessed December 26, 2012. http://www.bloomberg.com/news/2012-01-30/drugmakers-join-gates-foundation-in-fighting-tropical-diseases.html.

Tousignant, Noémi. "Broken Tempos: Of Means and Memory in a Senegalese University Laboratory." *Social Studies of Science* 43, no. 5 (2013): 729–53.

———. *Edges of Exposure: Toxicology and the Problem of Capacity in Postcolonial Senegal.* Durham, NC: Duke University Press, 2018.

Trapido, Stanley. "Imperialism, Settler Identities, and Colonial Capitalism." In *The Cambridge History of South Africa,* vol. 2, *1885–1994,* edited by Robert Ross, Anne Kelk Mager, and Bill Nasson, 66–101. Cambridge: Cambridge University Press, 2011.

Tsing, Anna Lowenhaupt. *Friction: An Ethnography of Global Connection.* Princeton, NJ: Princeton University Press, 2005.

UNAIDS. "GAP Report 2014." http://www.unaids.org/sites/default/files/media_asset/UNAIDS_Gap_report_en.pdf.

———. "South Africa HIV and AIDS Estimates (2014)." Accessed September 17, 2015. http://www.unaids.org/en/regionscountries/countries/southafrica.

Universities Allied for Essential Medicines. "2013 Annual Conference Guide." Accessed November 30, 2015. http://uaem.org/cms/assets/uploads/2013/09/2013UAEMConference_Packet.pdf.

van der Geest, Sjaak. "Anthropology and the Pharmaceutical Nexus." *Anthropological Quarterly* 79, no. 2 (2006): 303–14.

van der Geest, Sjaak, Susan Reynolds Whyte, and Anita Hardon. "The Anthropology of Pharmaceuticals: A Biographical Approach." *Annual Review of Anthropology* 25 (1996): 153–78.

Vaughan, Megan. *Curing Their Ills: Colonial Power and African Illness.* Redwood City, CA: Stanford University Press, 1991.

Velagaleti, Ranga, and Philip Burns. "A Review of the Industrial Ecology of Particulate Pharmaceuticals and Waste Management Approaches." *Particulate Science and Technology: An International Journal* 25, no. 2 (2007): 117–27.

Virk, Karen Poltis. "South Africa Today: Economical Development and Regulatory Standards Make the Country a Logical Choice for Trials." *Applied Clinical Trials,* November 1, 2009. Accessed October 5, 2017. http://www.appliedclinicaltrialsonline.com/south-africa-today.

Wainaina, Binyavanga. "How to Write about Africa." *Granta* 92 (2006). https://granta.com/how-to-write-about-africa/.

Waldby, Catherine. "Stem Cells, Tissue Cultures and the Production of Biovalue." *Health* 6, no. 3 (2002): 305–23.

Walwyn, David. "Determining Quantitative Targets for Public Funding of Tuberculosis Research and Development." *Health Research Policy and Systems* 11, no. 10 (2013): 1–8.

Watson-Verran, Helen, and David Turnbull. "Science and Other Knowledge Systems." In *Handbook of Science and Technology Studies*, edited by Sheila Jasanoff, Gerald E. Markle, James C. Peterson, and Trevor Pinch, 115–39. London: Sage, 1995.

Wendland, Claire L. "Opening Up the Black Box: Looking for a More Capacious Version of Capacity in Global Health Partnerships." *Canadian Journal of African Studies / Revue canadienne des études africaines* 50, no. 3 (2016): 415–35.

———. "Research, Therapy, and Bioethical Hegemony: The Controversy over Perinatal AZT Trials in Africa." *African Studies Review* 51, no. 3 (December 2008): 1–23.

WHO. "Waste from Healthcare Activities." Fact Sheet no. 253. November 2011. Accessed January 31, 2014. http://www.who.int/mediacentre/factsheets/fs253/en.

Whyte, Susan Reynolds, Sjaak van der Geest, and Anita Hardon. *Social Lives of Medicines*. Cambridge: Cambridge University Press, 2002.

Williams, Logan D. A. *Eradicating Blindness: Global Health Innovation from South Asia*. London: Palgrave Macmillan, 2019.

———. "Mapping Superpositionality in Global Ethnography." *Science, Technology, and Human Values* 43, no. 2 (2018): 198–223.

Williams, Sean. "7 Facts You Probably Don't Know about Big Pharma." *Motley Fool*, July 19, 2015. Accessed December 26, 2017. https://www.fool.com/investing/value/2015/07/19/7-facts-you-probably-dont-know-about-big-pharma.aspx.

Williams, Simon J., Paul Martin, and Jonathan Gabe. "The Pharmaceuticalisation of Society? A Framework for Analysis." *Sociology of Health and Illness* 33, no. 5 (2011): 710–25.

Wilson, Elizabeth A. "The Work of Antidepressants: Preliminary Notes on How to Build an Alliance between Feminism and Psychopharmacology." *BioSocieties* 1 (2006): 125–31.

Wisniak, Jaime. "The Development of Dynamite: From Braconnot to Nobel." *Educación química* 19, no. 1 (2008): 71–81.

Woodhouse, Edward, and Steve Breyman. "Green Chemistry as a Social Movement?" *Science, Technology, and Human Values* 30, no. 2 (2005): 199–222.

World Bank. *World Development Report 2013: Jobs*. Washington, DC: World Bank, 2012. https://doi.org/10.1596/978-0-8213-9575-2.

World Trade Organization. "Fiftieth Anniversary of the Multilateral Trading System." https://www.wto.org/english/thewto_e/minist_e/min96_e/chrono.htm.

Wright, Gwendolyn. *The Politics of Design in French Colonial Urbanism*. Chicago: University of Chicago Press, 1991.

Wynberg, Rachel, Doris Schroeder, and Roger Chennells, eds. *Indigenous Peoples, Consent and Benefit Sharing: Lessons from the San-Hoodia Case*. London: Springer, 2009.

Zweig, David, Chung Siu Fung, and Donglin Han. "Redefining the 'Brain Drain': China's Diaspora Option." *Science, Technology and Society* 13, no. 1 (2008): 1–33.

INDEX